From the Watching of Shadows

The Origins of Radiological Tomography

From the Watching of Shadows

The Origins of Radiological Tomography

Steve Webb

Senior Lecturer in Physics as Applied to Medicine,
Joint Department of Physics, Institute of Cancer Research,
University of London and Royal Marsden Hospital

CRC Press
Taylor & Francis Group
Boca Raton   London   New York

CRC Press is an imprint of the
Taylor & Francis Group, an **informa** business

First published 1990 by IOP Publishing Ltd.

Published 2019 by CRC Press
Taylor & Francis Group
6000 Broken Sound Parkway NW, Suite 300
Boca Raton, FL 33487-2742

ISBN 13: 978-0-85274-305-8 (hbk)

Visit the Taylor & Francis Web site at
http://www.taylorandfrancis.com

and the CRC Press Web site at
http://www.crcpress.com

British Library Cataloguing in Publication Data

Webb, Steve, *1948–*
  From the watching of shadows: the origins of radiological
  tomography.
  1. Tomography
  I. Title
  621.36′73

  ISBN 0-85274-305-X

Library of Congress Cataloging-in-Publication Data

Webb, Steve, Ph.D.
    From the watching of shadows: the origins of radiological
  tomography/Steve Webb.
        p.   cm.
  Includes bibliographical references.
  ISBN 0-85274-305-X
  1. Tomography—History.   I. Title.
RC78.7.T6W43   1990
616.07′57′09—dc20   89-26820
                                                          CIP

Typeset by BC Typesetting, Exeter EX2 8PN

Our science is from the watching of shadows . . .
Canto 85 in *The Cantos of Ezra Pound* 1956 (London: Faber & Faber and New York: New Directions)   © Ezra Pound, 1956

Contents

In which we discover numerous lone workers pioneering section imaging, find classic patents, uncover the acrimonious rivalry for priority of invention and sort out the horrendous terminology. Was the Frenchman Bocage really first or did the Italian Baese beat him to it? With the story of how tomography bounced around Europe before reaching America and Britain

2 The middle years, circa 1940–50

In which we learn how the War decade matured tomography and what Frank and
Takahashi were doing two decades too early heralding computed tomography

3 Circa 1950 and beyond

In which Watson's transverse axial tomography shows a clear lead over other
methods and Finnish teeth are sharply imaged. With reflections on what Watson
could do mechanically without electronics and an account of contemporary
production equipment with space-age names

PART 2 Modern History

4 Pioneers towards computed tomography

In which we salute two Nobel Prize winners and tell how with that glorious
wisdom of hindsight some earlier experiments may be perceived prophetic and
profound. An American neurologist has a train set, gramophone and old clock;
two Russians in Kiev bring Gabriel Frank's idea to life; astronomers, crystallographers and mathematicians rework Radon's treatise and pioneering nuclear
medicine tomographers were up to the same tricks

5 Patents for computed tomography 194

In which a dig around the patents shows there are more ways than one to skin the proverbial cat

6 Historical emission tomography 223

In which the scene largely shifts to America and the tale of section imaging of radioactive tracers is told. Mechanical movement, electronic photon detection and computational wizardry achieve what the X-ray tomographers of old could only dream of. There may seem to be little new under the sun

Preface

Imaging is now firmly established as a part of diagnostic medicine, and methods of creating images of sections of the body (tomography) are well developed. This book grew from rooting around to find the answers to questions such as the following, Who patented techniques for tomography? When was the first patent? Who built the first useful equipment? Did the early pioneers know about each other's work? Why were there so many different pieces of commercial apparatus? What was the relationship between the early inventors and the commercial companies? How did people go it alone and construct home-made equipment? What was the effect on section imaging of the two World Wars? As the study progressed, new questions emerged. Was there cross-fertilization of ideas from transmission to emission tomography or vice versa? Were digital techniques based on earlier analogue methods? When was the first X-ray computed tomography scanner constructed? What national differences influenced developments? What kind of people were responsible for these exciting developments? As one would expect, the well known names associated with the history of the physics of tomographic radiology are all here. However, a particular delight has been to be able to record the contributions of some less well known figures, who in a way have become personal heroes. Some of these were unable, because of the limitations of the technology of their day, to see their ideas fully evolve. As William Osler has written, 'In science the credit goes to the man who convinces the world, not to the man to whom the idea first occurs.' (Mackay 1977).

My fascination for some things of the past includes an interest in the origins of that broad class of radiological techniques generally referred to today as tomography, and I sense this is shared by others who, like myself, earn their living by developing and putting into practice modern techniques for medical imaging. As George Orwell wrote, 'Who controls the past controls the future.' (Mackay 1977).

There is a sense in which most scientific research is a continuum and that, however well prosecuted is each claim for novelty in development, there is generally a body of knowledge or experience lurking in the background on which the new work stands. This is not to say that authors fail to acknowledge this when communicating the results of their research, but nevertheless the extent of most reference to earlier work and

xi

other papers is usually limited to the more recent literature. More often than not the history is unknown to the new writer; it might be too old or a principle only known in another unrelated field, or simply 'lost'. The continuum is often only appreciated with hindsight. Much of the literature of tomography has this flavour and one intention of the present work is to draw together several of the disparate strands of the story of how body section imaging has come to be achieved.

Many connecting links between the well known landmarks may be appreciated. No value judgment in favour of the old is implied, quite the reverse. Whilst we today may marvel at what could be achieved in early times, no practising radiologist would claim that times were better then. Computed emission and transmission radiology are here to stay and have largely replaced older, mainly analogue methods, and rightly so. It has been recorded (see for example Weber 1973) that Newton wrote to Hooke, 'If I have seen a little further it is by standing on the shoulders of giants.' Likewise an appreciation of the role of the past in shaping present science has been an attribute of many eminent scientists, among them the pioneering medical physicist W V Mayneord. This approach was epitomized in his opening address to the 1950 International Congress of Radiology in London. Boag (1984) has written of Sylvanus Thompson, Founder and first President of the Röntgen Society, that he 'traced things back to their sources, referring to the pioneers and to the historical development of the subject from . . . wide reading and scholarly research'. Thompson was also responsible for the remark, 'Every scientist and his work is but a child of his scientific forefather.' (Etter 1965, Bleich 1960). I particularly like the quote from the German radiologist Rudolf Grashey (1876–1970): 'The technician wishes to, and the Röntgen physician should, find out about the transition from simple beginnings to the current perfected forms. He would then become more tolerant in the evaluation of the technical shortcomings discernible even today.' (Grigg 1965).

The process of extrapolating backwards can, however, be overdone. Many important discoveries can be seen with hindsight to have been precedented. Perhaps it is worth recalling that even the experiment which led Röntgen to discover X-rays (Glasser 1965) had been performed before but without the experimenters realizing the significance of what they saw. On 22 February 1890, Arthur Willis Goodspeed, Professor of Physics at the University of Pennsylvania, unwittingly made an X-ray image of two coins using a Crookes' tube. The plate was preserved by his coexperimenter, William N Jennings, as an unexplained curiosity, later to be exhibited to the American Philosophical Society on 21 February 1896 (Brecher and Brecher 1969). Goodspeed thus found a place in history as one of the men who *might* have discovered X-rays. In the period of fanatical nationalism in Germany immediately before and during the Second World War, Lenard's role in the run-up to the discovery of X-rays was greatly emphasized at a time when acknowledging Röntgen was out of favour (Etter 1965). There are indeed a number of developments which are extremely close to section imaging but which do not quite get there and the status of these is open to debate, as we shall see in this book.

This preface is intended in part as a vehicle for clarifying the scope of this work and to disassociate it from roles which were not my intention for this book. I have been conscious throughout of the dilemma of writing in the somewhat disparate styles which suit the historian of science and the scientist and must confess that where these were irreconcilable the latter was adopted. I want this to be a useful book, and without technical detail it cannot be so. We have all in our time reinvented wheels and the book is particularly directed to the physics student in the hope that this will be avoided and, moreover, that some cross-fertilization of ideas may prove beneficial for future work. Notwithstanding, where possible the approach has been lightened with quotations and, dare one admit, even anecdote. To convey some of the flavour of early times, illustrations from the earliest patents and papers have deliberately been chosen. Throughout there is comment and (hopefully informed) criticism. Again the tolerance of those irritated by anecdote is requested. The book is also for those who know some—hopefully not all—of these early applications of physics and engineering to the practice of diagnostic medicine. In this gathering together of stories is critique and comment and the author begs pardon of those wiser than he who may view these events differently. In order to remain as objective as possible, work is traced to secondary sources throughout and where possible substantiated with primary source material. Many people have been of assistance—listed later—and I thank them all.

I believe I have also succeeded despite the mathematical nature of the subject in keeping the number of equations in single figures. Inevitably the subject has been approached from the point of view of the physical scientist. However, here is physics underpinning medicine and this book is as much for all those diagnostic radiologists who enjoy the origins of their techniques. For the more general reader a glossary of technical terms is appended. Another appendix summarizes the important developments in radiology in general against which background tomography has evolved. Some readers may prefer to start there before commencing the main text. The work is restricted to the use of ionizing radiation to form images both by transmission and by emission. Imaging with other modalities, for example ultrasound, nuclear magnetic resonance and thermal energy, is excluded.

There has been no previous complete coverage of the physics of tomographic radiology from its earliest beginnings to modern times and this book aims to remedy this deficiency. As work progressed, questions began to frame themselves on a more detailed level and a guiding principle in writing became to try to balance attacking these questions in depth against the inevitable tedium of too much detail for the average reader. Of course no writer can expect to get to the bottom of all such questions and I very much welcome the reporting of errors of fact or inference. Neither was intended. A larger work would have done greater justice to some workers and to them I apologize.

Steve Webb
June 1989

Acknowledgments

I am grateful to all those who have replied to my letters soliciting primary source material. They are named individually in the references and with the captions for figures which they kindly supplied. I am particularly grateful for comments on parts of the manuscript by Professor B G Ziedses des Plantes, the pioneering tomographer, and to Professor C B A J Puijlaert (Utrecht) for introducing me. Mr E Mercer provided many personal reminiscences of William Watson. I am very grateful for sharing memories to Dr J Ambrose, Professor H H Barrett, Professor A B Brill, Professor G Brownell, Professor T Budinger, Dr S Derenzo, Dr M Goitein, Sir G N Hounsfield, Dr J Keyes, Dr D E Kuhl, the late Mr J B Massey, Professor R McWhirter, Professor G Muehllehner, Dr D R Pickens, Miss K Prior, Professor S Sakuma, Mr A B Strong, Professor M M Ter Pogossian and Professor Y L Yamamoto.

I should like to express my sincere gratitude to Mrs P Rumens, librarian at the Institute of Cancer Research, for helping me to obtain copies of oft-times obscure literature which our library did not hold. This task was crucial for the work and was undertaken willingly and enthusiastically. Her shared sense of delight when a search for a source was successful spurred on my efforts.

I should like to thank librarians and archivists Mrs B Weedon, Mr R M Jordan and the staff at the Library of the University of Utrecht, Queen Mary's Hospital, Roehampton, and The British Institute of Radiology. I am grateful for the encouragement of medical historians Professor W F Bynum and Dr J Guy. Assistance with scientific translation was provided by Mr I Adam, Dr M Blaszczyk, Dr S Cherry, Mrs A Edyvane and Dr A Nahum. I am grateful to Mr R Stuckey and Mr P Court for reproducing old photographs for this volume and to Dr R Bentley for providing word-processing facilities.

The source of each illustration is acknowledged in the figure captions. I am grateful to the publishers for permission to reproduce this copyright material. It proved impossible to trace all material to the rightful owners of copyright, particularly where journals ceased to be published early this century and where authors were thought to be deceased. For this, apologies are offered. I am, however, grateful to Mrs S Sugden and Mrs P Rumens for helping me to trace some very obscure publishers.

Dr J E Bateman read the whole of the manuscript and suggested many improvements which I am pleased to acknowledge. I am also grateful for comments on the manuscript by two Past Presidents of the British Institute of Radiology, Professors J W Boag and I Isherwood.

I should like to thank Sean Pidgeon (Commissioning Editor) and Jen Halford (Desk Editor) at Adam Hilger Publishers for their enthusiasm for and professional work on this project.

This book is dedicated to Linda, Tom and David.

List of tables

Part 1

Early History: Classical Section Imaging

Chapter 1

The Early Years,
1895–circa 1940

Modern diagnosis of disease makes a great deal of use of images of the patient, and 'medical imaging' is now well established. There can be few of us who have not at some time in our lives been subjected to a diagnostic procedure involving forming a picture of our internal anatomy or function. By far the most common procedure is the taking of a simple X-radiograph, perhaps to locate a suspected fracture or to assist with diagnosing a chest complaint. The use of ultrasound for imaging soft-tissue structure is also extremely common; its role in monitoring pregnancy is particularly well known.

In recent years, bold steps forward have been taken in harnessing new physical techniques for forming medical images and today we are in a position to create images probing the body with X-rays, with ultrasound, with radiofrequency pulses (nuclear magnetic resonance imaging) and with high-frequency electrical potentials as well as to create images from the photons which escape the body after a radioactive tracer has been injected (nuclear medicine). Images can also be created from naturally emitted thermal infrared radiation.

It is all too easy to forget that it is not yet a century since the only way to see inside the body was via invasive operation and that in the nineteenth century a doctor's knowledge of anatomy was obtained from anatomical drawings of dismembered cadavers. On 8 November 1895, Wilhelm Röntgen discovered the X-ray and heralded a new era for medical diagnosis. The development of diagnostic radiology was dramatically fast (see appendix 1), not surprising in view of the enormous benefit to both patient and doctor of non-invasive imaging. The other 'modalities' for making images are more recent developments largely belonging to the second half of the twentieth century and with respect to the most recent arrivals on the scene their roles are still being evaluated.

From its earliest beginnings, medical imaging has been an interdisciplinary activity in which both medical and non-medically qualified scientists have shared an interest. Their roles have been complementary; today imaging technology can be highly complicated and the physicist plays a central role. Manufacturing industry is vitally important. The radiographer who supervises the interaction between the patient and the imaging equipment and the radiologist whose responsibility it is to interpret the pictures today require an intimate knowledge of the physics and technology in order to carry out their tasks properly. As 'customers', the general public have come to have high expectations that their care will be improved with the coming of each new tool. Local newspapers carry stories of appeals for 'scanners' and most people are familiar with the idea that viewing the body as a series of isolated sections is inherently desirable. Indeed this revolution in radiology is acclaimed as a quantum jump in diagnostic improvement comparable only with Röntgen's original discovery.

Radiological science is rooted in physics and there have been many eloquent expositions of its early history. No doubt as the Röntgen centenary (1995) approaches there will be more. This book tells that part of the story which is concerned with the creation of body-section images using ionizing radiation from its earliest beginnings to the developments which can be regarded as the basis of modern imaging. Like all good stories in science it is the tale of individuals; their combined efforts underpin modern imaging science. Some names are well known to modern workers but a surprisingly large number of the early pioneers have been forgotten. If you are reading this as a radiologist, radiographer or physicist, challenge your colleagues to name who first suggested (and patented) the idea of forming cross-sectional images of patients. In fact this is not difficult to establish but much more complex detective work is required to unravel the full story of the events between 1920 and 1940. This is largely the story in chapter 1 where we shall meet each of the key players. For at least 10 years, perhaps a little more, a number of people laboured alone and with no knowledge of each other. In the mid-1930s the picture changed. As gradually they found out about each other, astonishing words appeared in print concerning claims for priority. It is all here in this chapter.

We shall in time come to see that a great variety of inventions were made with much the same aim, that of isolating in focus some particular plane in the patient in which the clinician was interested. The passage of time has perhaps blurred the distinctions between these and this is a shame because, once the barriers to communication were opened, there was a great deal of wrangling over which equipment was best. It is also fascinating to see what elaborate contraptions were designed and built. Hidden in these early developments were also the beginnings of modern 'computed tomography' (CT) and another theme through this book is to show the relation between CT and early section imaging. There are at least two (Frank and Takahashi) and possibly more pioneers whose work in the 1940s might be viewed as a direct precursor of CT. In the late 1960s a veritable crowd of workers were rushing along with contributory research. From the early 1950s, imaging in nuclear medicine was heading towards tomography and treading familiar paths of development similar to the earliest work with external X-rays. We shall tell the story in the last chapter when the interrelationship between section imaging in the two fields can best be appreciated.

Readers of books like this in my experience usually have different expectations. There will be those whose interest lies in the early engineering and with them in mind most of the illustrations have been taken from original patents and papers. These are not always the clearest expositions of the principles involved (though they should have been!) and so in places reasonably lengthy descriptions have been provided so that no confusion arises. Others may be more interested in the relationships between the developments and their originators, working conditions at the time, the way in which the information was disseminated and perhaps (the juicy bits!) the arguments between the pioneers. I have tried to satisfy them too and made liberal use of quotations from published work and from personal communication (with the originators of more recent techniques). Whilst personal comment from inventors is always interesting it is not always entirely objective and I like to think that the technology which we have in place today is a testimony to their combined efforts. The purpose of this book is to put these together for the first time and humbly to dedicate the result to them all.

1.1 Early section imaging

Modern medical imaging makes use of a large number of tomographic techniques by which physical properties of biological tissues are displayed on three-dimensional matrices of volume elements or voxels. Selected planes within the patient can be viewed by displaying the two-dimensional distribution when one of the three position coordinates is kept fixed. It is customary to regard such a plane as containing little contamination from any adjacent planes and modern reconstruction techniques ensure that this is by and large the case. Perhaps the most widely appreciated of such techniques is X-ray CT in which the biological property of interest is the linear attenuation of X-rays, which is principally related to anatomical structure rather than to biological function. Although the mathematical principles underlying the method were known early in the twentieth century, it was only when digital computers became available for implementing fast reconstruction algorithms that CT imaging became possible. The impact in medicine has been colossal and CT imaging is in widespread use.

Such has been this impact that it has tended to eclipse interest in earlier methods of producing clear images of selected planes within the body using X-rays. Many of these methods were quite ingenious and today are often referred to collectively under the generic title of 'classical tomography'. This cataloguing rather obscures the fact that early attempts to image the body in some kind of three-dimensional manner were many and varied. All kinds of generic title were proposed and adopted and claims for originality in this area were certainly as fierce as more modern rivalry in CT.

Part 1 of this book is an attempt to provide some perspective on early section imaging, to unravel the historical order of events and to reinstate some more detailed classification of the 'modalities' of classical section tomography. In part 2 the more modern history of CT is presented with the parallel development of emission tomography. If stereo X-ray imaging is included as a section-imaging attempt (the first papers were published in March 1896) the period of interest in classical tomography can be regarded as starting almost immediately after the discovery of X-rays by Röntgen in 1895 and flourishing in the period up to the beginning of the Second World War. Dead history, however, this is not, for classical tomography is still much in use, when CT would be uncalled for, and there is much to be gained from a study of early tomography in appreciating CT.

This was the 'steam age' of medical physics when things were often done very differently from today. A look at the early papers reveals the different approach. Many of the inventions were by clinical radiologists anxious to improve their practices. Many of the papers were from single authors, presumably lacking the team effort characterizing much modern research. Some of the papers have no list of references and appear to have been presented almost as the first word on the subject. Often the style of presentation was different, much use being made of the first person, and by today's standards some presentations were somewhat anecdotal. Nevertheless they lacked little in scientific approach although at least one 'famous method' was later discredited as strictly inaccurate. Above all, this was a golden age of 'in-house' construction. The development of early section imaging is also one dominated by geometry. No knowledge of advanced mathematics is needed to understand the basis of developments which were largely the result of applying carefully geometrical theorems to little more complicated arrangements than sets of similar triangles. Most workers got on with the job of building and testing equipment and did not go in for the elaborate simulations which are available through computers to us today. As a result it was not usually possible to deduce the effects of scattered X-rays on the images ahead of time.

Despite the apparent simplicity of the theory confusion seems to have been abundant. An American radiologist Moore (1939) who played an important role in getting section imaging established in the USA wrote, 'Body-section radiography is relatively new but an extensive literature already exists. This is strikingly complex and difficult to understand. Many writers on the subject, by whatever name they have chosen to call it, have described one or more somewhat related procedures under one heading in one article, thereby greatly confusing the reader. Understanding this subject is not made any easier by the conflicting often acrimonious claims made for priority.' Dewing (1962) wrote, 'There is no uncertainty about the physical basis of body-section röntgenography. . . . But almost everything else about it is very confused. The trouble is that the idea occurred separately to several people at about the same time. . . . These different investigators (about ten in all) were each fired with zeal and authorship. They put on an unseemly and even acrimonious wrangle over priority during the early 1930s.' We shall soon encounter (and need to have unravelled) the terms stratigraphy, planigraphy, tomography, laminography, planeography, serioscopy, antidiffusion stereography, stereoradiography, seriescopy, stereoröntgenometry, stereoröntgenoscopy, body-section radiography and many more! Perhaps the latter term 'body-section radiography', first introduced by Moore (1938), best sums up what this is all about. It has in fact a remarkably modern sound to it. However, as a collective word, the generic term 'tomography' is much used here to embrace all these, being based on the fundamental juxtaposition of Greek 'temnein' meaning to cut, and Greek 'graphia' meaning writing. The pioneering British radiologist Twining (1937a) preferred to adopt this term (see § 1.2.9) and it is of course the modern word for sectional imaging. This followed the adoption of the term officially by the International Commission on Radiological Units and Measurements (ICRU) in 1962 (see § 1.4).

The most obvious primary subdivision of techniques is into

(1) those by which a single X-ray image is made in which some particular plane is more in focus than the others—for this some movement of X-ray source and detector relative to the patient is required during the exposure (see § 1.2)—and

(2) those by which three-dimensional depth effects are created by viewing several radiographs made with the X-ray source in different locations relative to the patient, the exposures being taken only when the source had been repositioned—each contributing radiograph has no depth information (see § 1.3).

Today we would label the former as 'tomography' and the latter as 'stereoradiography'; in the period being reviewed, however, these techniques went under many names.

Table 1.1 summarizes some of the principal landmarks in classical tomography and stereoradiography and table 1.2 lists the principal patents. In this book the approach to the history of tomography is via the physics and engineering of the subject. A different historical perspective from a clinical reviewer has been provided by Westra (1973). We shall see in this chapter that the story of tomography unfolded in several geographically separated places, if not exactly simultaneously then at least for over a decade without any of the pioneers knowing of the others. For that reason any attempt to tell the story in strict chronological sequence has little meaning. Instead we shall meet each of the developments in turn and gradually the pieces of the jigsaw will fall into place.

Table 1.1 Some of the principal landmarks in classical tomography and stereo-radiography: the early years, to circa 1940.

Year	Worker; invention
	Tomography
1914	Mayer; moving-tube stationary-film method (not true tomography)
1915	Baese; Italian patent, 17 May 1915
1916	Baese; UK patent, application 16 May 1916
1917	Baese; UK Patent, accepted 19 July 1917
1917–21	Bocage; theoretical work
1921–2	Ziedses des Plantes; began work on 'planigraphy'
1921–2	Bocage; patent, application 3 June 1921, granted 4 May 1922
1921–2	Portes and Chausse; patent, application 3 October 1921, granted 3 August 1922
1927	Pohl; German patent, application 30 November 1927
1928–9	Kieffer; 'X-ray-focusing machine'
1929	Kieffer; patent, application 13 March 1929
1930	Vallebona; 'stratigraphy'
1930	Pohl; Swiss patent, application 22 December 1930
1930	Pohl; UK patent, application 24 December 1930
1931	Ziedses des Plantes; first paper on 'planigraphy'
1931	Bartelink; patent, application 6 June 1931
1931	Bartelink; first paper on 'planigraphy'
1932	Pohl; German patent, granted 28 January 1932
1932	Pohl; UK patent, granted 24 March 1932
1932	Pohl; Swiss patent, granted 30 June 1932
1932	Bartelink; patent, granted 15 July 1932
1934	Kieffer; patent on 'laminagraph', granted 10 April 1934
1934	Siemens and Reiniger; research planigraph

Table 1.1 (*continued*)

Year	Worker; invention
1934	Grossmann; French patent, application 6 August 1934, granted 18 October 1934
1935	Grossmann and Chaoul; 'tomography'
1936	Massiot; 'Planigraph'—the first commercial section imager
1936–8	Grossmann; USA patent, application 8 January 1936, granted 15 March 1938
1936	Andrews; historical review
1937	Massiot; the Biotome of Dr Bocage
1937	Twining; paper on 'tomography'
1936–8	Watson; UK patent on longitudinal-section apparatus, application 22 October 1936, granted 23 February 1938
1936–9	Vieten; patent on transverse tomography, application 12 March 1936, granted 3 March 1939
1937	Andrews and Stava; paper on 'planigraphy'
1937	Ponthus and Malvoisin; 'la stratiscopie'
1937–9	Watson; transverse axial tomography patent, application 29 December 1937, granted 29 June 1939
1938	Kieffer; paper on 'laminography'
1938	Keleket Company (USA); the commercial version of the Kieffer laminograph
1938	Bush; paper on 'tomography' (after Twining)
1938	Alexander; paper on planigraphy
1939	Moore; paper on body-section radiography and laminography (after Kieffer)
1939–40	Watson; US patent, application 7 September 1939, granted 9 April 1940
1940	Wheeler and Spencer; paper on 'planigraphy'
	Stereoradiography
1896	Elihu Thomson; invention of X-ray stereo imaging
1896	Mackenzie Davidson; description of stereo X-radiography
1911	Manges; paper on stereopelvimetry
1916	Mackenzie Davidson; book on stereo imaging
1935–6	Ziedses des Plantes; seriescopy patent, application 27 August 1935, granted 26 August 1936
1936	Kaufman; papers on planeography
1936	Massiot; visit to Ziedses des Plantes who requested to copy the Seriescope
1938	Cottenot; paper on serioscopy
1939	Chausse; paper on stereoradiography
1940	Kaufman and Koster; paper on serioscopy

Table 1.2 Major patents before 1940 (in chronological order of application).

Date Registered	Date Granted	Author	Country	Number
16 May 1916	19 July 1917	Baese	UK	100491
3 June 1921	4 May 1922	Bocage	France	536464
3 October 1921	3 August 1922	Portes and Chausse	France	541914
30 November 1927	28 January 1932	Pohl	Germany	544200
13 March 1929	10 April 1934	Kieffer	USA	1954321
22 December 1930	30 June 1932	Pohl	Switzerland	155613
24 December 1930	24 March 1932	Pohl	UK	369662
6 June 1931	15 July 1932	Bartelink	Switzerland	155930
6 August 1934	18 October 1934	Grossmann	France	771887
4 November 1934	17 May 1939	Sanitas	Germany	676594
30 March 1935	14 July 1938	Sanitas	Germany	663476
27 August 1935	26 August 1936	Ziedses des Plantes	UK	487389
7 September 1935	1 September 1938	Sanitas	Germany	665336
8 January 1936	15 March 1938	Grossmann	USA	2110954
12 March 1936	3 March 1939	Vieten	Germany	672518
21 March 1936	7 December 1939	Both	Germany	686022
22 October 1936	23 February 1938	Watson	UK	480459
29 December 1937	29 June 1939	Watson	UK	508381
7 September 1939	9 April 1940	Watson	USA	2196618

1.2 Single-film slice focusing

In 1937, two Americans summarized the state of the art and their work is often quoted as a seminal clarification of the developments in the early years. The importance of the work of Andrews and Stava (1937) lies in the systematic mathematical analysis presented which distinguished between planigraphy (due to Bocage and Ziedses des Plantes), tomography (due to Grossmann) and stratigraphy (due to Vallebona). These four, all working independently played a major role in the story of body-section imaging. The fundamental differences between the techniques which they developed can be seen from what follows. In all methods *relative movement* between X-ray source, patient and detector is the key to isolating some particular region 'in focus' at the expense of blurring others.

1.2.1 Stratigraphy and Vallebona

In *stratigraphy* the X-ray source and film detector (cassette) are fixed to a pendulum such that they are perpendicular to the pendulum and parallel to each other whatever the orientation of the arrangement. Both tube and detector describe arcs. This experimental arrangement is generally attributed to work by Vallebona (1930a–f, 1931) (born 1899) (figure 1.1) in Italy. Alessandro Vallebona was born in Genoa on 2 March 1899. He

Figure 1.1 Alessandro Vallebona, pioneering Italian tomographer. Photograph taken on the occasion of the Second International Congress of Radiology in Stockholm, 23–27 July 1928. (From Renander (1928).)

11

graduated from the University of Genoa in 1923 and obtained a degree in radiology in 1930. He also took a degree in physical therapy in 1935 and became chief of the Department of Radiology and Physical Therapy at Genoa City Hospitals in 1938. In 1944 he became Professor of Radiology and Director of the Institute of Radiology of the University of Genoa. He was awarded the French Légion d'Honneur (Bruwer 1964).

In Vallebona's stratigraphy the intention is that a plane normal to the upright pendulum shall remain in focus (figure 1.2). That this is not the case was simply shown mathematically by Andrews and Stava. If the pendulum moves in one plane, then only the *line* of axis of rotation remains in focus. If the pendulum moves in a circular or spiral fashion about a point, then only that *point* remains in focus. This method is characterized by the fact that the central axis of the X-ray beam does *not* lie normal to the tomographic plane (except at just one orientation), instead pointing at the axis of rotation. In his earlier paper, Andrews (1936) does not appear to have yet come to the conclusion that this method is invalid. He wrote, 'This is a very simple and practical method.'

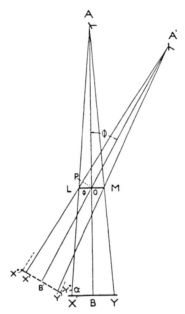

Figure 1.2 The principle of section imaging by stratigraphy. The X-ray source moves from A to A' and the detector correspondingly moves such that the source and detector form a rigidly coupled pendulum pivoting at O. The intention is for plane L–M to remain in focus but in reality only a line normal to the plane of the diagram and through O is sharply imaged. (From Andrews and Stava (1937). Courtesy of the American Röntgen Ray Society.)

On the other hand, Grossmann (1935a) stated that 'stratigraphy (does) not fulfil the main condition of giving a truthful reproduction of a plane'. Vallebona's earliest stratigraphic apparatus (also strictly invalid) was a complementary design (figure 1.3). The X-ray tube and detector remained fixed whilst the object was rotated about a pivot thought to have defined the in-focus plane. Vallebona used the equipment to make planigrams of skulls. Gebauer (1956) records that Vallebona appreciated the critics of his first and second methods and agreed the sections were less useful than those obtained with other methods. Methods later developed by Vallebona with Bozzetti improved on the earlier stratigraphic techniques.

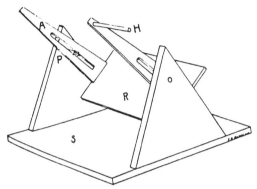

Figure 1.3 A schematic drawing of Vallebona's original stratigraphic apparatus. The film was placed on the table S and the object to be investigated on the rocking plate R. The arms A could be adjusted to bring the pivotal line P to any position within the body. The source of X-rays (not shown) was positioned vertically above the apparatus which was rocked via a handle H. The apparatus was used to make section images of skulls. (From Andrews (1936). Courtesy of the American Röntgen Ray Society.)

Vallebona (1931) published in English his technique for stratigraphy (not calling it that in this paper) in a short and somewhat anecdotal paper in 1931. There is much use of the first person and the paper reads as a casual commentary on thoughts going through his mind. It is almost as if we stumble by accident on these glimmers of the new technique. The paper is only two pages long and has no diagrams, tables, references nor results. Indeed the paper begins by discussing magnification radiography before changing topic to 'a technical method for the radiographic dissociation of the shadow'. To get some of the flavour of the presentation it is worth quoting verbatim what Vallebona wrote: '. . . it would be most advantageous to have an image, so to speak, distinct from that of all the various planes and without the image of them being superposed on that one of interest to us. In other words it would be an advantage to do with X-rays what we do with the microscope; that is, to focus only one plane of the object. If we could adopt with Röntgen rays lenses and reflectors as we can with light rays, or if we could put only one plane in focus as with the microscope, the problem would be very simple; but, since it is impossible to do this at present, it is necessary to try some other way . . . and avoid—up to a certain point, at least—the superimposition of extraneous shadows. These . . . procedures should be studied further and perfected, for there is no doubt but that they can render useful service in advancing the improvement of radiographic technic.'

In a later paper, Vallebona (1933) seems set on putting right the historical perspective of stratigraphy and placing his own contributions firmly in context. This paper refers only to the work of Bartelink and Ziedses des Plantes although, as we shall soon see, there were by 1933 far more workers in body-section imaging than just these two and Vallebona. He opens by referring to two papers in the April 1933 *Fortschritte auf dem Gebiete der Röntgenstrahlen* by Bartelink (1933) and by Zeidses des Plantes (1933) and summarizes the theory of section imaging. Vallebona later refers to another paper by Ziedses des Plantes (1932) and it would appear that this was the first he knew of the work of Ziedses des Plantes. After reading this, Vallebona (1932a) announced his own claim to work in this field in the official journal of the Italian Radiological Society. He says he came to the conclusion that the methods of Ziedses des Plantes (see §1.2.3) were no different from those of Bartelink and himself. In fact we now know that this was not so. It was indeed true that Vallebona had developed stratigraphy and published a little earlier (in February 1930) than Bartelink. He referred to six papers (Vallebona 1930a–f) on technical methods of body-section imaging which he had invented. Vallebona's (1933) paper is a repetition of what he said in these six papers, illustrated with photographs of the various apparatuses which he invented. These were all based on the pendulum principle of stratigraphy with different designs and various combinations of relative movement between patient, source and detector. It would appear that Vallebona's interest was in laying claim to having invented body-section imaging and that at the time of writing he was unaware of other work in the 1920s, such as that of Bocage, Portes and Chausse, and Pohl. Incidentally it is in this paper that Vallebona says the name 'stratigraphy' was coined by Professor Busi in a discussion of his conference paper (Vallebona 1930d). This is also referred to in the review by Ott (1935). Vallebona (1950b) returns to the question of the origins of body-section imaging in a paper on the much later technology of transverse-section laminography (see §1.2.11 and chapter 3). He wrote, 'In 1930 I succeeded in obtaining the first laminographs. I was followed in rapid succession by Ziedses des Plantes (1931), by Bartelink (1932a), by Grossmann and by Chaoul (1935a–c). Since then the method had been adopted under the names of planigraphy, stratigraphy, lamin-

14

ography and tomography. Preparatory experimental work tending toward cross-sectional study of the body was made by Kieffer in 1938.' Once again the record seems accurate but somewhat incomplete as will be more apparent when in this chapter the contributions from these and other workers will have been reviewed *in toto*. It might appear (Vallebona 1930d) as if Vallebona's work goes back as far as 1928 when he was first investigating microradiography.

In § 1.2.3 we shall look in detail at the work of Ziedses des Plantes. There was a great deal of correspondence in the press between this worker and Vallebona, largely concerning the claims for originating planigraphy (see, for example, Vallebona 1932b, 1934, Ziedses des Plantes 1934c). Other interesting papers include those by Bozzetti (1935) and Vallebona and Bistolfi (1935a–c) who elaborated on stratigraphic apparatus following in the mould of Vallebona. A specialist journal *Stratigraphia* directed by Vallebona was begun in 1956 (Grigg 1965). Vallebona may not have been first but his work spanned decades and was of seminal importance.

1.2.2 Planigraphy and Bocage

In *planigraphy* the X-ray tube and the detector move in horizontal and

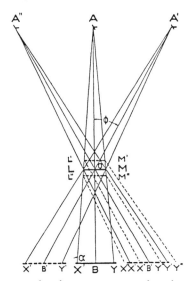

Figure 1.4 The principle of section imaging by planigraphy. The X-ray source moves from A' to A" through A whilst the detector follows an antiparallel path at a constant separation distance from the plane of the source movement. All points in the plane through LOM remain in focus whilst other points in the volume are blurred. The motion shown is linear but other movements were engineered including spiral, circular, epicycloidal and square. (From Andrews and Stava (1937). Courtesy of the American Röntgen Ray Society.)

parallel planes and describe either linear, circular or spiral movements or combinations of such movements such that the source and detector execute coupled antiparallel motions (figure 1.4). By this arrangement all points within the plane parallel to source and detector and passing through O remain in focus. All other points in the volume (not in this plane) are blurred. It can be shown that this result is analytically exact (Andrews 1944). The tube and cassette holder require to remain in equidistant parallel planes.

The planigraphic methods of body-section röntgenography are attributed (Andrews 1936) to Bocage (1921–2) (figure 1.5) who first described the principles in *French Patent Application 536464* (figure 1.6(*a*)) on 3 June 1921, in Paris at 3.22 pm in the afternoon! The patent was published on 4 May 1922, the year before that of Röntgen's death (Glasser 1938). André Bocage (1892–1953) was a dermatologist who invented the method in 1916 during military service in the First World War when he was in charge of the X-ray examination of wounded soldiers. Bocage was born in Paris of parents who had come from Lorraine (JAMA 1965). He entered medical school in 1909 and rose to become physician to the Welfare Centre of the Préfecture of the Seine and Chief Consultant in Dermatosyphilology at the Pasteur Institute. He served in

Figure 1.5 The father of body-section imaging, André Edmond Marie Bocage. (From Bruwer (1964) *Classic Descriptions in Diagnostic Radiology* vol 2. Courtesy of Charles C Thomas, Publisher, Springfield, Illinois.)

RÉPUBLIQUE FRANÇAISE.

OFFICE NATIONAL DE LA PROPRIÉTÉ INDUSTRIELLE.

BREVET D'INVENTION.

XII. — Instruments de précision, électricité.

2. — Appareils de physique et de chimie, optique, acoustique.

N° 536.464

Procédé et dispositifs de radiographie sur plaque en mouvement.

M. André-Edmond-Marie BOCAGE résidant en France (Seine).

Demandé le 3 juin 1921, à 15ʰ 22ᵐ, à Paris.

(a)

Délivré le 13 février 1922. — Publié le 4 mai 1922.

DEUTSCHES REICH

AUSGEGEBEN AM
15. FEBRUAR 1932

REICHSPATENTAMT

PATENTSCHRIFT

№ 544 200

KLASSE **30**a GRUPPE 6

P 56609 IX/30a

Tag der Bekanntmachung über die Erteilung des Patents: 28. Januar 1932

Ernst Pohl in Kiel

Verfahren und Vorrichtung zur röntgenphotographischen Wiedergabe eines
Körperschnittes unter Ausschluß von davor und dahinter liegenden Teilen

(b)

Patentiert im Deutschen Reiche vom 30. November 1927 ab

RÉPUBLIQUE FRANÇAISE.

OFFICE NATIONAL DE LA PROPRIÉTÉ INDUSTRIELLE.

BREVET D'INVENTION.

XII. -- Instruments de précision, électricité.

2. — Appareils de physique et de chimie, optique, acoustique.

N° 541.914

Procédé pour la mise au point radiologique sur un plan sécant d'un
solide, ainsi que pour la concentration sur une zone déterminée d'une
action radiothérapeutique maximum, et dispositifs en permettant la
réalisation.

MM. Félix PORTES et Maurice CHAUSSE résidant en France (Hérault).

Demandé le 3 octobre 1921, à 16ʰ 4ᵐ, à Paris.

(c)

Délivré le 9 mai 1922. — Publié le 3 août 1922.

Figure 1.6 The title pages of the three historic patents by Bocage, by
Pohl and by Portes and Chausse. (Reproduced with permission.)

both World Wars, being severely invalided in the Second World War (Bruwer 1964). His interest in radiology developed during the First World War when he asked to be assigned to a radiologic unit. Following the Armistice he returned to Paris and worked with Pierre Marie, Weill and others, graduating third in his class and taking an *MD Thesis* which won him the silver medal.

Bocage's career was interrupted by the Second World War. He was severely wounded on 22 May 1940 in the face of the German advance on his hospital which had been evacuated. It was not, however, until some years later that his health began to decline and he was able to work until the end of 1952. He then retired to his house in the country (Loiret) until he died on 10 July 1953 (Hillemand 1953).

The patent in 1921 is widely credited as the first ever patent in body-section imaging and Bocage's theories were of seminal importance. In the patent, Bocage proposed three separate planimetric procedures (cited by Ott (1935)). The central ray from the X-ray tube is always normal to the tomographic plane whatever the position of the tube and irrespective of whether the motions were linear, circular, square or Archimedean spirals. This is not a strict requirement and later workers suggested that the central ray could be directed at a fixed point. However, Andrews and Stava (1937) concluded that the necessity to pivot the tube and cassette in some kind of cradle, whilst ensuring the central ray follows AOB (in figure 1.4) in all positions would require a complicated mechanism. Grossmann (1936–8) later did achieve this (see § 1.2.6).

There can be no doubt of this very early exposition of the principles of slice focusing in the patent by Bocage. He wrote (my translation)†, 'The X-ray tube focus (and) . . . the (detector) plate . . . are linked by a mechanism such that elementary displacements are always synchronous, parallel, of opposite direction and in a relationship of constant dimensions. . . . Under these conditions there exists in the space between them a single fixed plane in which each point has always for its shadow the same point on the film, such that only the organs contained in this plane have a sharp image. The other organs only give diffuse shadows able to change partially the intensity of the original image but without adding new lines.' A translation of the whole patent has been made by Bricker and published by Bruwer (1964) in the book of reprints of classic papers in diagnostic radiology.

Bocage's patent is not lengthy and he did not propose nor construct equipment in the 1920s, it not being until 1938 that he played a further active role in section imaging. In reviewing the work of others we shall

---

† A great deal of the early literature was published in French, Dutch and German; it is hoped that factual information has been correctly translated but nuances and emphases resulting from style may have been overlooked.

have recourse many times to refer back to the work of André Edmond Marie Bocage as the father of section imaging.

1.2.3 Planigraphy and Ziedses des Plantes

Bocage's patent contains only schematic drawings of how source and detector movements may be properly executed. There are no engineering drawings although the schematic drawing does include provision for antiscatter grids. It would seem that Portes and Chausse (1921–2) described similar methods some 4 months later (see §1.2.4) and Pohl (1927–32) also described the principles in a German patent (see §1.2.5). Andrews (1936) is very clear on what he believes is the historical order of events. Ziedses des Plantes stated in his *Doctoral Thesis* (1934a) and also in a private communication to Andrews that he invented the method independently in 1921–2 (figure 1.7) but he did not publish his experimental results until a series of papers beginning in 1931 (Ziedses des Plantes (1931 *et seq*)). He was subsequently informed in 1933 that Bocage had taken out a patent in 1921 and Pohl in 1927 but that both

Figure 1.7 The pioneer of experimental planigraphy, Bernard Ziedses des Plantes, taken at the time when he invented body-section planigraphy. (From Littleton (1976) *Tomography: Physical Principles and Clinical Applications* The Williams and Wilkins Company. Reproduced with permission.)

patents had not led to practical use. He was also informed of the work of Vallebona. There is thus no doubt that all these other workers together with Portes and Chausse put the principle on record before Ziedses des Plantes and the reviewers of the 1930s are undecided whether to afford to Ziedses des Plantes the accolade of having invented planigraphy and hence of having been the forefather of modern tomography. They are agreed, however, that his thesis was the most elaborate and complete of all workers in this field. Grossmann (1935a) also noted the fact that many workers appeared to have found independent solutions to the same problem. He referred to them as 'partially identical'.

Whatever the disputes may be concerning the claims for priority, there can be no doubt of the fundamental importance of the work of Ziedses des Plantes and, if Bocage is afforded the honour of being recognized for depositing the first patent, Ziedses des Plantes is often accoladed as the pioneering experimenter. Westra (1972) clarified the events in the Laudation on the appointment of Professor Ziedses des Plantes as Honorary Member of the Netherlands Society for Radiology. He wrote that Ziedses des Plantes submitted his idea to a röntgenologist in 1921 but was told that the method would be of no use. It was not until 1928 that he returned to the idea and conducted simple experiments. Some of these involved radiographing hemiskulls placed on a gramophone turn-table irradiated from a stationary source. The section resting on the turntable was viewed sharply whilst other sections were blurred by the spiral movement. It would appear that manufacturers also dissuaded him by leading him to believe that the method was of no practical value and also (see § 1.4) pointing to earlier patents. Ziedses des Plantes' first paper (1931) appeared just 1 month before the meeting in November 1931 at which he and Bartelink coincidentally presented similar inventions.

The method of Ziedses des Plantes differs from those of some other workers (and in particular those of Bocage) in that the central ray always points at the same location on the film as described above with reference to figure 1.4. The X-ray source and the detector are constrained to parallel equidistant planes. Figure 1.8 shows a schematic drawing of the equip-ment developed by Ziedses des Plantes and figure 1.9 is a photograph of the equipment. Figure 1.10 clarifies the engineering, being a diagrammatic form of equipment developed by the Massiot Company in the mid-1930s based on Ziedses des Plantes' ideas (see also figure 1.11 and § 1.2.8). Referring to figure 1.8, when the tube R moves with the support arm A, the movement is conveyed to the film F by lever H pivoting at S (which defines the in-focus plane). The equipment was set up to execute circular movements of the tube and detector, the power being applied by hand rotation of an arm attached to the detector holder (see figure 1.10). By joining this arm to a fixed peg on the ground via a cord, spiral movements could be executed and Ziedses des Plantes was an advocate

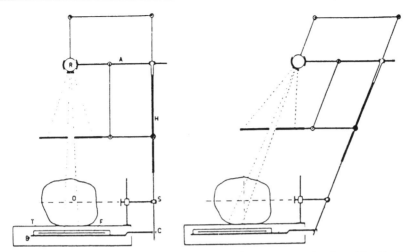

Figure 1.8 A schematic diagram of the planigraphic apparatus developed by Ziedses des Plantes; for description see text. The diagram shows two positions which the system might take up during linear or spiral planigraphy. The plane OS remains in focus. (From Ziedses des Plantes (1932).)

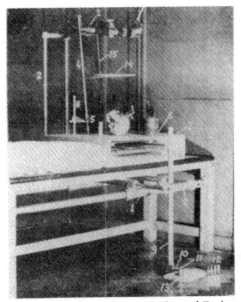

Figure 1.9 A photograph from the *Doctoral Thesis* of Ziedses des Plantes showing his prototype apparatus for planigraphy. The X-ray tube 3 can be moved together with the support 2 in which case the bar 9 of the support 2 is held fast. A lever 4 transfers the movement of the support 2 to the carrier 6 on which the cassette is placed. A cord 10 is wound around the peg 11, as the support 2 turns, thus producing a spiral movement. A skull may be seen on the table. A schematic diagram of equipment based on this prototype is shown in figure 1.10. (From Ziedses des Plantes (1934a).)

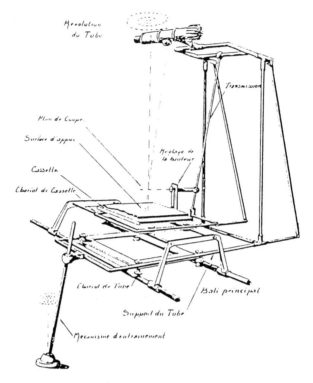

Figure 1.10 A schematic diagram of the planigraphic apparatus developed in the 1930s by Massiot et Cie and based on the concepts of Ziedses des Plantes, apparatus pictured in figure 1.9. The mechanism for producing the spiral movement is very clear. The actual Massiot apparatus is shown in figure 1.11. (From Massiot (1936).)

of the potential for improved blurring by this arrangement. By varying the size of the pegs, different spiral movements were achievable. He used the equipment to study the anatomy of the skull. Ziedses des Plantes (1932) published the drawings of this experimental arrangement in *Acta Radiologica* with descriptions in the German language. The firm of Siemens–Reiniger constructed a research model planigraph in January 1934.

Ziedses des Plantes is also credited with the invention of the device, still used today, to calibrate the plane of cut (in-focus plane). This comprised a wooden cylinder (today Perspex is used) in which are embedded lead numerals at their numerical distance in cm from one end together with lead shot spaced at 1 mm intervals. This device was placed on end on the table at the same time as the clinical exposure whereupon the numerals and shot in focus indicate the exactly defined tomographic plane. This invention was patented by Ziedses des Plantes in 1936.

Figure 1.11 The planigraphic apparatus designed by Massiot and tested by Naud, based on the principles developed experimentally by Ziedses des Plantes. The schematic corresponding to this is to be found in figure 1.10. (From Naud (1936).)

It is interesting to note how, in these early days, workers often published the same work several times in different journals and in different languages. Ziedses des Plantes (1934b) published the essence of his method in French, in a very short paper in which he directed the reader to his Dutch (Ziedses des Plantes 1931) and German (Ziedses des Plantes 1932) papers for fuller details. In this 1934 practical note we find one of the clearest explanations of how Ziedses des Plantes came upon the method. He stated, 'I found the method in 1921–2. At that time I was in my first year medical studies at the University of Utrecht after having studied at the Technical University of Delft. We were occupied with much microscopy from which was born the ideas underlying my work.' This paper showed the technique diagrammatically, also the numerical geometric registration system and some results from imaging dry skulls. He also showed how several parallel planigrams at different depths in the patient could be obtained simultaneously by placing several parallel spaced-out films in the moving film carriage, an idea which was re-proposed later by Watson (1937–9) for transverse axial tomography. As a postscript to his 1934 paper, Ziedses des Plantes explained how, after submitting his work for publication, he was told of Bocage's patent. He pleaded total ignorance, said Bocage's patent had not to his knowledge been implemented practically and that he (Ziedses des Plantes) was the sole originator of his invention. Perhaps because he indeed performed the earliest experimental work he is today often quoted as the inventor of classical section imaging (figure 1.12). Bocage, however, documented the

Figure 1.12 The pioneer of experimental planigraphy, Bernard Ziedses des Plantes. (From Bruwer (1964) *Classic Descriptions in Diagnostic Radiology* vol 2. Courtesy of Charles C Thomas, Publisher, Springfield, Illinois.)

earliest methods. The work of Ziedses des Plantes interested and was reviewed (adversely) by Van der Plaats (1932) who achieved a notable career in medicine, radiology and radiotherapy. Van der Plaats (born 1888) also trained in Utrecht (Bruwer 1964) as well as in Batavia, Frankfurt and Vienna. He must also be one of the few radiologists to have represented his country in the Olympic Games (for fencing in Berlin in 1936). Further biographical information on Professor Bernard Ziedses des Plantes is to be found in § 1.4.

1.2.4 *Planigraphy and Portes and Chausse*

Two other Frenchmen were hot on the heels of Bocage as originators of planigraphy. The *French Patent* 541914 of Felix Portes and Maurice

Chausse was delivered on 3 October 1921 at 4.04 pm in Paris and published on 3 August 1922 (figure 1.6(c)) (Portes and Chausse 1921–2). Figure 1.13 is reproduced from this patent. The principle by which a single plane may be seen sharply is clearly described. The curve a represents the plane trajectory of an X-ray tube c. The film follows an appropriately magnified or minified trajectory vertically displaced from the plane of a and parallel to it (for example curve b or b'). Another parallel plane containing o and o' defines the constant factor between the radii of these two trajectories such that any point within this plane such as o and o' remains in focus on the film. For example c connects via o to t (or via o' to t') and o and o' act as the vertices to conically generated surfaces. Portes and Chausse write (translated from French), 'The point c represents the emission of X-rays, the plane oo' . . . o^p is the plane being studied in the solid, the plane bb' . . . b^p is that of the sensitive receiving screen. It is so easy to work out that the projected image is such that the points in the plane oo' . . . o^p are seen clearly, whereas those of anterior and posterior planes are blurred. As a consequence, a device embodying the conditions described above resolves the problem of the X-ray imaging of a plane in a solid to the exclusion of others.' (For example points such as m, not in the plane oo' . . . o^p project to a circle of different radius at the detector and hence are blurred when the radius of rotation of the detector does not match.)

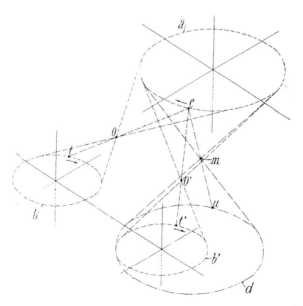

Figure 1.13 A schematic diagram of the planigraphic technique from the historic patent by Portes and Chausse (1921–2). The diagram illustrates circular planigraphy and is explained in detail in the text.

Diagrams were drawn to show possible ways in which the above geometrical movements could be realized experimentally. In particular the circular movement of the detector was drawn as being obtained from two orthogonal sinusoidal motions with the film sliding in a frame which itself slides in a fixed orthogonal framework. Portes and Chausse, however, claim that 'the equipment which has already been schematically described can admit numerous variations more particularly (*a*) on the method of executing the displacements of the X-ray tube, (*b*) on the mechanism of transmitting the movement to the detector screen and (*c*) on the position of the plane in focus'. They thus laid claim essentially to all planigraphy embodiments, although circular trajectories were the example used. They go on to offer further possibilities, showing that not only can a series of planes parallel to some starting plane be sequentially imaged but also that, by rotating the solid, other oblique planes cutting the solid may be examined.

The fifth and last figure in the patent (figure 1.14) shows a detailed mechanical arrangement of plates, shafts, wheels, ball joints, gears and handles all of which are arranged to provide the appropriate movements. There is absolutely no doubt that, unlike the patent of Bocage which, although earlier, had no engineering drawings, this patent clearly showed exactly how to build a planigraphic apparatus. The question of who invented section imaging seems to reduce to the two patented claims of Bocage and of Portes and Chausse. The former was registered earlier but appears substantially less well advanced.

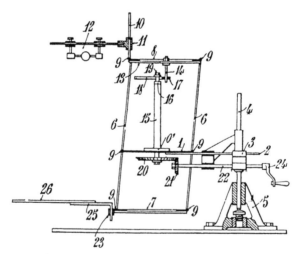

Figure 1.14 The engineering drawing from the patent by Portes and Chausse (1921–2) describing planigraphy. Turning the handle 24 causes the spindle 14 to describe a circle, taking with it the system of levers which cause the X-ray tube 12 and detector 26 to execute coupled circular movements.

Finally Portes and Chausse explain how the movements intended for planigraphy identically enable radiotherapy to be better localized if instead of a diagnostic X-ray tube a therapy source is used—essentially very early treatment planning of dynamic therapy! They wrote, 'The same device permits localization via a determined aim of the therapeutic action of penetrating radiation. In effect by limiting the size of the beam by an appropriate shutter one can (1) avoid exposing adjacent tissues to the (always dangerous) prolonged irradiation, (2) concentrate the active beam on the various stages of the neoplasm or cancer being treated, (3) prolong the therapeutic action whilst reducing to a minimum the chances of a lesion occurring in healthy tissue and (4) shorten the interval between radiotherapy treatment fractions.'

Whether the equipment was ever built, used for radiodiagnosis or radiotherapy is not known but clearly this was a historic patent in that the principles it pioneered are essentially those of modern tomography and multiple-beam radiotherapy.

1.2.5 Planigraphy and Pohl

The 1920s were years of rarified activity in section imaging. Now we meet another lone pioneer. The *German Patent* 544200 (figure 1.6(*b*)) of the German manufacturer Ernst Pohl (1878–1962) (figure 1.15) is dated 28 January 1932 with the date of application 30 November 1927 (Pohl

Figure 1.15 The industrialist Ernst Pohl from Kiel. Photograph taken on the occasion of the Second International Congress of Radiology in Stockholm, 23–27 July 1928. (From Renander (1928).)

1927–32). It was entitled 'Method and device for X-ray photographic reproduction of a body cross-section under conditions excluding structure from in front and behind'. There is no doubt from this patent not only that Pohl had designed an ingenious apparatus for planigraphy but also that he was able to describe alternative implementations. The patent includes drawings of methods in which

(a) by moving the X-ray source and film relative to a stationary body or

(b) by moving both the body and the film relative to a stationary X-ray source or

(c) by moving both source and body relative to a stationary film, planigraphy was achievable.

In all three geometries proposed, the central ray of the X-ray tube points at a fixed location in the in-focus plane. The preferred embodiment of Pohl's ideas is reproduced in figure 1.16, taken from his patent. The X-ray source 1 is supported within a cradle 7. The cradle is pivoted on a bearing 9 which is rigidly mounted onto a vane 3 which can rotate on an axis 2 and wheel 11 relative to a fixed baseplate 21. The rotation is provided by motor 10 driving a belt 23. The angle of the cradle to the axis of rotation is variable via screws 20 in slots 19. The mechanism is counterweighted at 17. The cradle supports at its upper part a plate 8 (which could be a film or a screen—see later). The body 14 is supported by the couch 13. A tie rod 4 couples the rotational motion to a second wheel 12 on the same drive belt via connectors at 6, 7 and 9.

The principle of operation is as follows. The aim is to maintain all points in the plane 16–15–16 in focus whilst all other points not in this plane are blurred. The motor drives the vane 3 through 180°. The assembly 7 is constrained to follow the vane, guided by the tie rod from the other wheel 12. In moving through 180° the cradle is pulled from a position in which the X-ray tube is at 1 to that in which the X-ray tube (broken lines) is at 1a. The cradle itself does not rotate about itself but nutates about the axis defined by 9. The planes of rotation of the X-ray tube and a plate at 8 are throughout parallel and equidistant. It can be seen that, with these movements, points such as 16 remain in focus throughout at the plate 8 where 16 can represent any location in the plane containing 15. Other points not in this plane are blurred. The blurring is of course circular with this apparatus rather than linear. The plane of section (in-focus plane) can be varied by adjusting the tilt of the cradle relative to the axis via the screws 20 in slots 19. It is clear that, if the plate 8 were a film, the desired aim would be achieved with hard copy. It was alternatively proposed to place a lens directly above a screen at 8 in which the tomographic image could be viewed. Pohl's apparatus was clearly a feasible design which would achieve the desired objective. We have

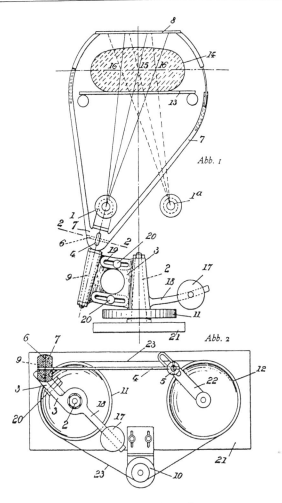

Figure 1.16 The engineering drawing from the German patent by Pohl (1927–32) showing in detail an apparatus by which planigraphy may be achieved. The apparatus is described in the text.

unfortunately no evidence to suggest it was ever constructed and Andrews (1936) reports that Ziedses des Plantes was told that it had not been developed. Ziedses des Plantes (1988a) has recently confirmed this and the same view is held by Littleton (1976).

Ziedses des Plantes (1932) refers to the work of Pohl. He wrote that he was of the opinion that his thoughts were entirely new and would appear to have been in the process of negotiating for a patent when he learnt of Pohl's 'Prinzippatent'. He affirms that his invention was independently conceived.

Pohl (1930–32a) also took out a patent in Switzerland (*Swiss Patent* 155613, application on 22 December 1930 and granted 30 June 1932) with a similar title. The details described in this patent are different although based on identical principles. Pohl appears to have given his earlier ideas more substance as can easily be seen from figure 1.17, taken from this patent. The X-ray tube and detector execute coupled circular trajectories with a constant geometrical ratio of radii. They are coupled via

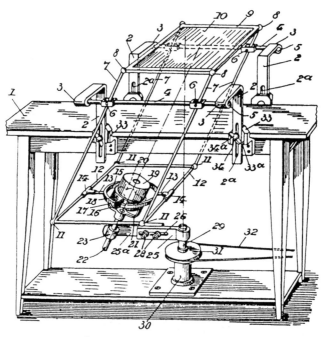

Figure 1.17 An alternative apparatus for planigraphy (to that in figure 1.16) from the Swiss patent by Pohl (1930–2a). See text for description.

a multijointed (11) eight-sided framework 7,12 which pivots about the plane of the section of interest (defined by pivots 6). The height of this plane above the table top is adjustable as is the geometry for rotation. This improved apparatus is driven from a single belt drive and has provision for a film detector 10. Once again the central ray points at a fixed location. It would appear that the detector could also be a fluoroscopic screen with an optical viewing system as in the German patent (not shown in figure 1.17). The Swiss patent also shows alternative experimental arrangements. When the detector was a fluoroscopic screen and the movements were synchronized circular trajectories, the images of different layers in the patient would appear to move in circles of different radii on the screen. A double convex lens was arranged in such a way that the radius of its circular rotation could be varied. In this way a particular plane in the patient could be selected for viewing as the images from this plane would appear stationary when the rotations of the image from this plane and that of the lens were concentric. Pohl (1930–32b) also patented this apparatus in the UK. It was not until the 1950s that Watson (1953b) was able to implement these ideas experimentally (see §3.3).

1.2.6 Tomography and Grossmann

The set of classes of section-imaging techniques is completed by tomography which is always associated with the name of Grossmann. In *tomography*, a superficially complicated arrangement (figure 1.18) also leads to sharp plane imaging. The X-ray tube and detector are once again attached to a rigid pendulum pivoting at O but in such a manner that,

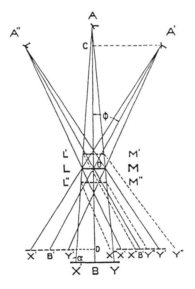

Figure 1.18 The principle of section imaging by tomography. The X-ray source moves from A′ through A to A″ with its central ray always pointing at the pivotal point O. The detector (at the opposite end of the pendulum connecting to the X-ray source) is constrained to be parallel to the sharply focused plane LM. The figure shows rays by which it is clear how all points (for example L and M) remain sharply focused. Points not in the plane LM are blurred. The pendulum swings through an angle 2ϕ during the exposure. The method by which the detector is constrained to the horizontal may best be appreciated by looking at the figures of Grossmann's apparatus. (From Andrews and Stava (1937). Courtesy of the American Röntgen Ray Society.)

regardless of the orientation of the pendulum, the detector always remains parallel to the plane of the cut. The central X-ray beam is always directed at a single fixed point in the plane of section. This arrangement is different from that of Vallebona because of the important change regarding the orientation of the detector (which gives the method its validity). Once again all points in the parallel plane passing through O remain exactly in focus whilst other points are blurred. One practical disadvantage of tomography is that the movement of the system must be linear rather than circular or spiral; hence blurring is only possible in one direction. In clinical practice the direction was chosen normal to the direction of principal features, for example longitudinally to blur the ribs in chest radiographs. Compared with the planigraph, however, the practical mechanics were thought to be simpler. The three classes of section imaging are summarized in table 1.3.

Table 1.3 Summary of the classes of section-imaging (slice-focusing) techniques. (See also Watson (1940) for a more detailed family classification.)

Class	Arrangement	Motion
Stratigraphy	The X-ray tube and cassette are perpendicular to a rigid pivoting pendulum	Linear, circular or spiral
Planigraphy	The X-ray tube and cassette move in parallel equidistant planes with reciprocal motions	Linear, circular or spiral
Tomography	The X-ray tube and cassette are attached to a rigid pendulum but the detector is always parallel to the tomographic plane	Linear

Tomography was invented by Grossmann (1935a) (born 1878) (figure 1.19) who claimed the following advantages over the other techniques.

(a) Circular and spiral planigraphy require excessive dose (3–15 times that of non-tomographic exposure). Pendulum tomography can be achieved with no time and dose penalty. He disagreed with some workers' reasoning for long trajectories for the focal spot which was based on the requirement for maximal blurring of small-scale features. His disagreement rested on the premise that in practice it was large-scale features which were to be blurred; to support the argument, he produced detailed calculations of the blurring factors for different structures under differing trajectories.

Figure 1.19 Gustave Grossmann, Director of Siemens–Reiniger–Veifa GmbH and Director of Siemens and Halske AG, Berlin. Photograph taken on the occasion of the Second International Congress of Radiology in Stockholm, 23–27 July 1928. (From Renander (1928).)

(b) The equipment of Ziedses des Plantes demanded complicated tube motions of translation and oscillation. Grossmann reported that the apparatus of Ziedses des Plantes was never used clinically although the evidence in Ziedses des Plantes' (1934a) thesis shows this to be an inaccurate remark.

(c) Exact correspondence of movements between source and detector is difficult to achieve with complex motions. Exact adjustment is difficult.

Moore (1938) disagreed with Grossmann's view that a linear movement was sufficient. Grossmann's tomographic equipment is illustrated in figure 1.20. The important feature which made this arrangement different from the second method of Bocage and the stratigraphic pendulum of Vallebona was the system labelled VZLY. V is a sliding frame which enables the plane of cut to be selected. An arm L connects the film tray B to V via pivots at Z and Y. The film tray is pivoted at C to the pendulum. This parallelogram arrangement ensures that the cassette is always parallel to the plane eAe whatever the orientation of the pendulum. Grossmann provides a clear description of why it is desirable in planigraphy and is automatically assured in pendulum tomography that the central beam of the X-ray unit points at a fixed location. This is not a mathematical

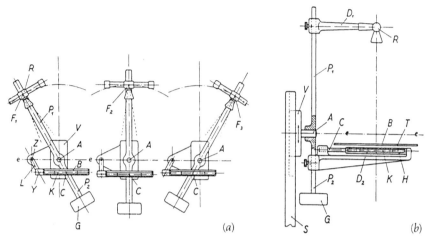

Figure 1.20 Schematic diagrams showing the apparatus of Grossmann for tomography. (*a*) Profile view showing the pendulum movement. (From Grossmann (1935a).) (*b*) Side view. (From Andrews (1936). See text for explanation. Courtesy of the American Röntgen Ray Society.)

requirement for such methods to work (the only strict requirement is that the source and detector execute coupled synchronous motions) but without the condition the field of view is too small. The argument is reproduced diagrammatically in figure 1.21. Grossmann's apparatus was manufactured by the firm of Sanitas, Berlin, and is shown in figure 1.22. According to Grigg (1965) the Sanitas Tomograph was first manufactured in 1933 and was a design much copied. The firm ceased trading early in 1960. Clinical work on lung tomography performed in Berlin was reported by Chaoul (1935a–c), by Greineder (1935) and by Chaoul and Greineder (1936). Henry Chaoul (figure 1.23) was Professor of Radiology in Berlin. He was born in Egypt in 1883 and after the Second World War returned to settle in Alexandria. Grigg (1965) noted that he was subsequently treated for radiation damage in Munich. He died in 1964. Kurt Greineder (born 1906) was a Lebanese radiologist working in Beirut. The first worker to make use of Grossmann's Tomograph in the UK was McDougall who had installed equipment at the Preston Hall British Legion Village (McDougall 1936, Kerley 1937). The inspiration came from a visit to Berlin (Rowbottom and Susskind 1984) (see also §2.2). A paper by Gassul (1936), reproducing a photograph of the Grossmann Tomograph and a very convincing diagram showing how tomography works, appeared with a lengthy Russian text showing the dissemination of the method beyond Europe.

Grossmann (1935b) described the experimental details of tomography and its clinical application in great detail and a companion paper

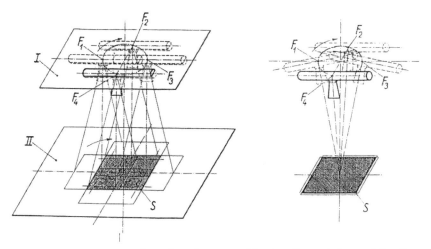

Figure 1.21 Diagrams to show the advantage of having the central ray of the X-ray tube permanently directed at the pivotal point in section imaging. On the left the source always points vertically downwards and circulates via $F_1F_2F_3F_4$. Only a small part of the body section S is permanently exposed (the centre of the cross visible in the shaded section). On the right the circulating tube is supported by Cardanic bearings and so guided that the central ray is permanently directed to the centre of the layer to be radiographed. The whole area of S is permanently exposed. (From Grossmann (1935a).)

Figure 1.22 The Grossmann Tomograph. (From McDougall (1940).)

Figure 1.23 Henri Chaoul, Director of the Röntgen-Institutes der chirurgischen Universitätsklinik der Charité, Berlin. Photograph taken on the occasion of the Second International Congress of Radiology in Stockholm, 23–27 July 1928. (From Renander (1928).)

(Grossman 1935c) elaborated on theoretical considerations. These papers place tomography in context with developments by earlier workers whom Grossmann acknowledges. Much of the work in these two German papers is similar to that reported in English in 1935 (Grossmann 1935a).

Grossmann (1934) patented his invention in 1934 in France. A year before his English paper explaining why he thought tomography was superior to planigraphy this French text reveals these ideas. Unlike the earlier patents cited in this chapter this patent begins by reviewing the known procedures for planigraphy. There is no feeling that this is from another lone worker claiming to have invented section imaging. He does not specifically name the earlier workers but the problems with planigraphy are systematically pointed out by way of prefacing his invention. He refers to the long arcs of travel in planigraphy, the small field of view consequent on the X-ray tube pointing vertically, the potential difficulties of construction without vibration and so on. The small field of view is, he says, particularly a problem with the then modern concept of shielded tubes with collimation.

The equipment described by Grossmann in the patent utilized such a shielded tube, a detector which remained parallel to the section to be imaged and the pendulum arrangement described above. The arc of travel was large (45°) for maximum blurring. A grid against secondary radiation was provided, focused in the direction orthogonal to the transverse movement. The tomographic principle was clearly explained with equations including describing how the equipment could be set up optimally for each study. In the actual embodiment of the principles described in the patent (figure 1.24), Grossmann employed a neat arrangement for varying the source to detector distance. This was achieved by turning a screw thread 9 which passed through the support for the detector 7 and also through a counterweight 8. However, 7 and 8 were reamed with screw threads of opposite sense and hence moved in opposing directions when the thread was turned. By arranging for the counterweight to be equal to that of the detector carrier with its working parts, the pendulum was arranged to be in equilibrium with its centre of gravity at the pivot for all possible source-to-detector locations. The whole equipment could be raised or lowered relative to the patient by the screw thread 16 in order to select different planes within the patient for imaging. The mechanism for ensuring that the detector and grid planes remained parallel to the plane of section incorporated rods 17 and 19 with pivotal points 18 and 20. In view of the requirement for the distance of the detection plane from the axis of rotation to be variable, the rods 19 were telescopic. It was also intended that, as the level of the plane of interest, defined by the position of the pivotal point, was lowered in the

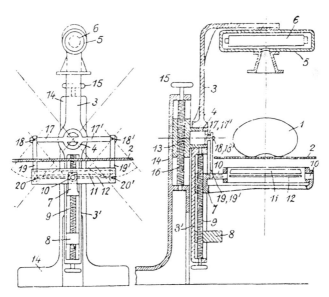

Figure 1.24 The patented drawings of the Grossmann Tomograph. See text for description. (From Grossmann (1934).)

body, the detector was correspondingly raised by the same amount in order to maintain the disposition of the detector in relation to the table. Grossmann also provided for the equipment to be electrically driven with the firing of the X-ray tube synchronized to the oscillation of the pendulum. Comment was also made on the need to orient the plane of swing at 45° to dominant directions of body structures such as ribs.

1934 would appear to be the date after which it was becoming clear that a number of workers had described inventions with much the same aim. (Interestingly not a mention of tomography is to be found in Glasser's (1933) classic text on the origins and development of radiology.) It is significant that Grossmann's own patent was the first to set his work in context with that of others. From this year on, a number of papers appeared reviewing these inventions and what one might call 'derivative workers' set out to repeat and refine the methods. One cannot escape noticing also that all this early development in section imaging had taken place in Europe outside the UK. With the exception of Baese's (1915–7) UK patent (see § 1.2.10)—and even Baese's invention was Italian—and Pohl's UK patent, all the patents were published in languages other than English, mostly French and German. Most of the pre-1934 papers were also not in the English language with the exception of that from Vallebona (1931).

Ott (1935) provided the most comprehensive German review. He began by recalling the firm basis from the theoretical work of Bocage. He reviewed the contributions from Portes and Chausse, from Pohl and from Vallebona, ascribing to the latter the honour of having published the first clinical images around 1930. He outlined the contributions from Bartelink and Ziedses des Plantes indicating how their work, together with that of Grossmann, became the basis of experimental prototypes built commercially by Siemens–Reineger and Sanitas. It would appear that Ott played a leading role in persuading these companies and installed prototype equipment for evaluation at his Institute.

Naud (1936) also shows figures of apparatus developed by Bozzetti (1935) based on similar principles to Grossmann's machine (figure 1.25). Either the tube could remain stationary and both the patient and the film rotate about a pivot (figure 1.25(*a*)) or the arrangement could be as in figure 1.25(*b*). This was enough to guarantee a plane in focus and was a fundamental improvement on the stratigraphic technique of Vallebona. A similar arrangement was engineered by Jones and Bradley-Bowron (1939) (see § 1.2.9 and figure 1.41).

Ott's conference review was followed by papers by Chaoul (1935c), Grossmann (1935e) and Vogt (1935) which also concentrated on the new work with the Tomograph. The ensuing discussion of these four papers included comment by Ziedses des Plantes and by Bartelink (see § 1.2.7).

In the USA, Grossmann (1936–8) patented apparatus performing linear

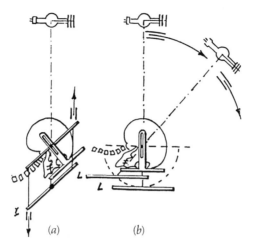

Figure 1.25 Bozzetti's apparatus for tomography. The detector is shown at L. The plane in focus is shown as a broken line (through the skull) in (*b*). (From Bozzetti (1935).)

planigraphy. This is shown in figure 1.26. The X-ray tube 22 and a trolley 4 (which houses the film detector) were mounted on rails on which they could travel linearly. The two linear movements were connected by a system of levers 62, 66, 70, 44, 40, 42, etc. The lever 38 pivots about a point in the plane ab, thus defining this as the in-focus plane. There is a telescopic arrangement at 60 to take account of the fact that the levers 40 and 38 are not of constant length when a pendulum motion is not used. The X-ray tube was arranged to pivot about a bearing 24 such that the central axis of the rays always pointed at a fixed location M in the plane ab. It is apparent that this is an acceptable arrangement for generating sections. The moving parts are supported by arrangements of pillars such as 14, and the patient is arranged to lie between them on a table whose height may be varied to give different sections. The method is strictly 'planigraphy' because source and detector remain equidistant but the rotation of the source to ensure the central ray passing through M gives this the same field-of-view advantage as is usually associated with pendulum tomography.

The tomographic technique which is identified with the name of Grossmann exemplified the efficacy of one-dimensional movement for blurring out-of-focus planes leaving one plane sharply in focus. Some writers, for example Griesbach and Kemper (1955), have classified the one-dimensional blurring methods into groups depending on whether the X-ray source executed an arc of a circle (tomography) or a straight line (linear planigraphy). As we have seen, Grossmann patented an example of both. The firm Sanitas held many patents including three (Sanitas 1934–9,

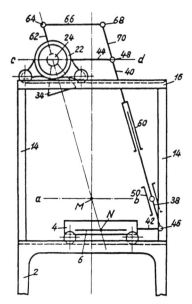

Figure 1.26 A different arrangement for linear planigraphy (but with the source pivoted to give a large field of view) patented by Grossmann in the USA. See text for details. (From Grossmann (1936–8).)

1935–8a, 1935–8b) where the source executed an arc of a circle as in the Grossmann (German) patent. An example of the other subclass was the patent by Both (1936–9) in which the source and detector were constrained to move in parallel straight-line one-dimensional paths.

1.2.7 Planigraphy and Bartelink

In referring to the work of Bartelink, Andrews (1936) claims that his technique was analogous to that of Vallebona in using a fixed pendulum and arranging for the X-ray tube and detector to face each other with the central ray passing through the pivot. Andrews thus links Bartelink to the concept of stratigraphy (see §1.2.1). Bartelink's (1931–2) patent, however, is a claim to an invention for planigraphy and the principles described therein (in German) are quite different. Bartelink's patent shows that he too was a contender for the invention of planigraphy. It is interesting to note that Pohl's Swiss patent was in the process of examination at very much the same time as that of Bartelink. Bartelink lodged his application on 6 June 1931, and Pohl on 22 December 1930. Pohl's patent was granted on 30 June 1932 only 15 days before Bartelink's patent was approved (15 July 1932). No one appears to have noticed the striking similarity between the two!

Bartelink begins his application with pre-amble explaining that a secure diagnosis cannot be obtained from X-ray images in view of the blurring of overlying and underlying structures. He claims the need for section imaging and proposes an invention to effect this improvement. The mathematical principles of planigraphy were written down and diagrams provided to explain how points in a preferred plane remain in focus as the X-ray source and detector execute a coupled motion in opposing directions. The choice of such a plane is made by varying the pivotal point of a system of levers which couple the linear motions.

Figure 1.27 is a detailed engineering drawing extracted from the patent. The planigraphic rather than stratigraphic principle is apparent. 1 is a supportive framework. The X-ray tube 5 rests on a framework 4 which can slide horizontally relative to a groundplate 3. The film or imaging plane is supported on the framework 6 which is in a plane parallel to the plane of movement of the X-ray source. The 'trailers' or chassis supporting tube and detector are coupled via four rods 7 which are arranged so their ends pass through holes drilled in these trailers. The rods 7 themselves pivot in ballraces 8, connected to the frame 1. The altitude

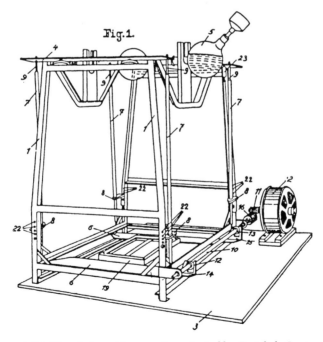

Figure 1.27 The planigraphic apparatus patented by Bartelink. See text for description. Notice the cams 12 and 13 which convert rotation of the spindle 10 into reciprocating linear movement of detector and X-ray source. (From Bartelink (1931–2).)

41

of the pivotal points above the groundplate can be varied via the locating holes 22. Such an arrangement enables the source and detector to move in coupled antiparallel proportional motion. Bartelink provides a novel motive power for the planigraphic motion which distinguishes this equipment from competitors. The motor 2 turns a spindle 10 via gears 11 and 16. The spindle has rigidly attached a pair of eccentric cams 12 and 13 which rest in slots 14 and 15 attached to the film carrier. Hence as the spindle rotates, the film carrier—and hence also the X-ray tube carrier—executes a sinusoidal motion. The patent does not suggest values for the amplitude of this motion but it would appear that, as the pivotal point is much closer to the detector than to the tube, the detector plane has a small oscillation amplitude whilst the X-ray travel is of large amplitude. It is also not stated how many oscillations take place for each exposure but the implication is that there would be more than one. The long lever effect of moving a large mass of X-ray tube is also claimed to act as a damping device.

Finally in Bartelink's patent he provides a schematic drawing and description of how circular or spiral planigraphy could be achieved. This would appear to be an addendum to the patent as he does not give it the same weight in that no engineering drawings were filed. It is interesting to speculate whether he was aware of the theoretical work of others, specifically of Bocage who first explained this principle.

Bartelink (1933) described his equipment for sectional X-radiography (planigraphy) and its use in clinical practice. He also explained how the equipment could be used to produce stereoröntgenograms. Some insight into the time scale of his work is given by this paper. In it he wrote that it was 2 years before his engineer could produce usable apparatus in finished form and that the first human application was in March 1930. Vallebona (1933) confirms this date for Bartelink's work. This being the case it would seem that Bartelink made his first contributions to planigraphy in 1928 or even earlier.

Andrews' (1936) review contains an error which causes some confusion about the developments made by Bartelink. He wrote that on 2 November 1931 the Dutch–Canadian Bartelink (1932a) (born 1894) demonstrated results obtained with stratigraphic apparatus at a meeting of the Dutch Association of Electrology and Röntgenology in Amsterdam. Andrews identified the technique as stratigraphy similar to that of Vallebona. Unlike Vallebona's equipment in which the height of the pivot of the pendulum could be varied in relation to the object to redefine the stratigraphic plane, in Bartelink's equipment the pivot remained fixed but the height of the object on the table could be varied. He referred to two papers by Bartelink (1932a, 1933) but the titles of these two papers have been switched in his list of references by mistake. Also the figure of Bartelink's apparatus which Andrews reproduces and ascribes to the 1932 paper would appear to be a redrawing with minor alterations of the figure in the 1933 paper which depicts apparatus not for stratigraphy but for planigraphy. In § 1.4 this correct perspective on Bartelink's work has been supported recently by Ziedses des Plantes.

Like many others who published their method in several languages, Bartelink (1932b) published his planigraphic method in French. In this short paper he acknowledges the possibility of getting better diagnostic information by stereography but says the technique is unpopular in some quarters. He explains the principles of planigraphy very clearly, in particular pointing out that, although the mathematics allows a single section to be precisely focused, in practice penumbral effects will lead to a thickening of the region in focus and that it would be more proper to refer to a layer rather than a plane in focus. The paper appears to be a conference report because he refers to showing illustrations (which are not printed) and to running out of time. For some reason he is referred to as Barteluck in this paper.

Bartelink in Nijmegen thus claimed apparatus for planigraphy. We have already referred to the work of Bocage, Ziedses des Plantes, Portes, Chausse and Pohl as pioneers for planigraphy. In §§ 1.2.8, 1.2.10 and 1.2.12 we shall discover other workers (Ponthus, Malvoisin, Kieffer and Baese) whose work also fits into this pattern and shall return to the question of claims for honours.

1.2.8 Planigraphy, Ponthus and Malvoisin and other derivative mid-1930s French workers including the Messrs Massiot

The mid-1930s was a period of feverish activity in section imaging, particularly in France where the method originated in two of the earliest patents. Ponthus and Malvoisin (1937a) reviewed the contributions of the early workers in body-section imaging. As with most of the reviews in the 1930s, only a selected set of workers was mentioned. They recalled the work of Bocage, Vallebona, Chaoul, Grossmann and Ziedses des Plantes and other important workers whom we shall meet later, notably Mayer and Baese; they specifically make mention of the different terms used to describe techniques, preferring themselves the term 'stratigraphy'. Homing in on the work of Bocage, they reiterated that the precise curve followed by the X-ray tube can be of various types: circular, spiral or curvilinear. They referred to the thesis of Naud (1936) also reviewing techniques. Ponthus and Malvoisin themselves adopted one of the planigraphic techniques proposed by Bocage, Portes and Chause, Pohl and also by Bartelink, namely that in which the X-ray source executed a circular trajectory in view of speed and continuity. They explain the principle by which a single plane remains in focus whilst other planes are blurred. Despite the fact that this is the principle of planigraphy they call this stratigraphy. Their apparatus, however, comprised a quite novel feature. Instead of mechanically providing for the synchronous circular movement of the X-ray detector (as in the drawings of Bocage) the X-ray image was arranged to fall on a stationary fluorescent screen. A pair of mirrors, parallel to each other and inclined at 45° to the axis of rotation, were arranged to rotate in a plane parallel to that of the source and in synchronism with it. By this arrangement a virtual image of the in-focus plane was caused to be formed at a fixed stationary location. The principle of the method is shown in figure 1.28.

Ponthus and Malvoisin demonstrated the effectiveness of this technique by constructing an optical model in which the X-ray source was replaced by an optical source; the phantom examined comprised three semitransparent plates at different depths and optical shadows were cast

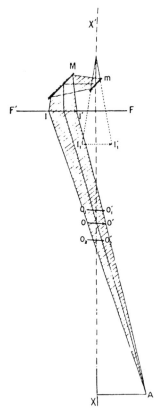

Figure 1.28 The principle of planigraphy as exemplified with the novel arrangement by Ponthus and Malvoisin. Mirrors M and m at 45° to the plane of section rotate in synchrony with the X-ray source A and generate a virtual image at I_1I_1' when viewed from X. The plane OO' remains in focus. FF' is a fluoroscopic screen. (From Ponthus and Malvoisin (1937a).)

on a screen. The light source and the system of mirrors were rotated by pulleys operated by hand. The apparatus is illustrated schematically in figure 1.29. When the pulleys were not operated, a superposed image of all three planes was seen by the viewer; when the pulleys were operated to provide a rotation of 3 or 4 rev s^{-1}, only the central plane remained stationary in focus whilst the images of the other planes were blurred. It appears that the depth resolution was some 3–5 mm.

Ponthus and Malvoisin had presented the essence of 'la stratiscopie' in an earlier paper the same year (Ponthus and Malvoisin 1937b) in which they outlined the principles of operation of the optical analogue. This paper is notable for the light it sheds on other Frenchmen who were reviewing the methods of section imaging in the 1930s, namely Massiot (1936), Naud (1936), Buffé (1936) and Belot (1937). Naud was a student

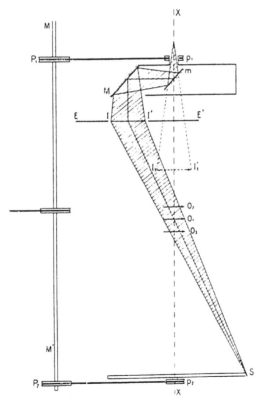

Figure 1.29 The schematic diagram of the optical analogue system built by Ponthus and Malvoisin. A light source at S shines on semitransparent plates O_1, O_2 and O_3. An image is produced on a screen EE' which when viewed from X through rotating mirrors M and m yields a virtual image at I_1I_1'. Provided that the light source and the mirrors rotate in synchrony, only one plane is sharply focused. (From Ponthus and Malvoisin (1937a).)

of Buffé in Lyon. Ponthus and Malvoisin (1937b) wrote, 'This present note . . . does not discuss the study of the value of different methods of sectional radiography. It is known nevertheless that the usefulness of the method began to be perceived in France, particularly in the domain of pulmonary radiology and during the second international conference organized in Strasbourg by Gunsett (born 1876) (figure 1.30) for the study of lung radiography.' They referred to the earlier work of others on stratigraphy, planigraphy and tomography and then specifically went on to explain that stratigraphy is not so useful because of the impossibility of *a priori* localizing the layer of interest. Although they recognized methods of empirically determining the depth of section, they concurred with Thiel and with Belot (1937) (born 1876) that such techniques were costly and rather unscientific.

Figure 1.30 Auguste Gunsett, Director of the Anticancer Centre of Strasbourg. Photograph taken on the occasion of the Second International Congress of Radiology in Stockholm, 23–27 July 1928. (From Renander (1928).)

The justification for their development of 'la stratiscopie' would appear to be the solution of the problem of locating the coordinates of the plane of section. Whatever the perceived limitations of the other techniques might have been, the designs from Ponthus and Malvoisin were intriguing. We might wonder whether an X-ray system was ever built?

The conference held in Strasbourg on 15 November 1936, organized by Gunsett, would appear to have been an important landmark in the history of body-section imaging. Belot (1937) (figure 1.31) has reviewed

Figure 1.31 J Belot, Radiologist of the Hôpital St Louis, Paris. Photograph taken on the occasion of the Second International Congress of Radiology in Stockholm, 23–27 July 1928. (From Renander (1928).)

the conference. Upwards of a hundred international delegates were assembled from France, Germany, Italy, Switzerland and Holland. The UK is not mentioned. There were clinicians and radiological technicians and inventors. Gunsett reported that there had been many important technological improvements in radiology since an earlier congress, 4 years previously. Many workers presented their findings that pulmonary section imaging was proving useful, notably Reissner from Stuttgart, Camino from Cambo-les-Bains and Schmidt from Heidelberg-Rohrbach. It seems that clinical section imaging was well established by 1936 although it was remarked in several places that the techniques were costly and had to be reserved for specialized institutes.

The report by Belot is rather vague in places, not distinguishing clearly between different techniques and it is thus a little difficult to determine which were well established in 1936. One thing is clear. Vallebona was at the conference expounding the merits of stratigraphy and in particular there was discussion of the most appropriate name to apply to the method. He preferred 'stratigraphy' in view of the finite stratum examined. Thiel preferred the term 'radiotomie' in lieu of tomography. He utilized a stereographic technique to determine the plane of interest prior to section imaging to avoid, as he put it, 'fishing around in the lungs'. Massiot showed a copy of Bocage's French patent and commented that it pre-dated the techniques used by Ziedses des Plantes, Chaoul,

Grossmann and Vallebona. He went to some lengths to explain the principles of selectively focusing one plane with sharp detail, other planes being blurred. Here then is another part of the jigsaw in which it was becoming clear that the mid-1930s was the time when several workers were realizing the similarity of several independently discovered techniques for body-section imaging. Grossmann was also commenting on this in 1934 (see § 1.2.6) and Andrews in 1936 (see § 1.2.2).

The paper by Massiot (1936) to which Ponthus and Malvoisin refer is very interesting and is another mid-1930s review of body sectioning. Ponthus had invited Massiot to speak in Lyon to the Société d'Electroradiologie. He had accepted particularly since he had strong familial links with Lyon. He wrote, 'It is because of the memory of my grandfather and of the personal merit of my father that I must be amongst you.' His father (G Massiot (1875–1962)) had been a pioneer of X-ray equipment production with many achievements to his name (see, for example, Grigg 1965). He had two sons, Marcel (1902–46) and Jean (born 1901) both of whom directed the X-ray equipment manufacturing interests of the Massiot Company at various times. On this occasion, Jean Massiot wrote that he was embarrassed not to have new equipment of his own to present and decided instead to review 'analytic radiography'. He had been to visit Ziedses des Plantes in Utrecht who had given him several documents enabling comparisons of techniques to be made. Massiot quickly dismissed stereo imaging as requiring considerable imagination. He referred to Bocage's (1921–2) patent saying that 'circumstances unhappily did not permit Dr Bocage to realize his projects'. He then went on to report how Ziedses des Plantes rediscovered 'in all good faith' the same invention after an interval of several years, this time with the merit of actually building an apparatus. Note that Massiot's statement that Ziedses des Plantes' invention was 'several years' after Bocage's patent (that is after 1921–2) does not accord with Ziedses des Plantes' own account (see Ziedses des Plantes' (1934b) paper and § 1.2.3). The record has been put straight by Ziedses des Plantes (1988a) (see § 1.4), dating his work also to 1921–2.

Massiot (1936) carefully classified the known body-section techniques into 'three big families'—stratigraphy of Vallebona, tomography of Chaoul and Grossmann and planigraphy of Bocage and subsequently Ziedses des Plantes. Into his group of explanatory diagrams he put a figure of his own apparatus for tomography which he says was designed by his father and announced in June 1935. Preliminary apparatus for studying pendulum movement was built at l'Hôpital de la Pitié and was being used by M le Médecin Colonel Buffé and his student Dr Naud. Massiot and his colleagues had always thought, like Chaoul and Grossmann, that spiral planigraphy was overcomplicated. In reality they found the experimental evidence showed tomography also had its difficulties and they illustrated with cases from Ziedses des Plantes how spiral tomography was essential for certain clinical cases. Also, by using non-screen–film in a multilayered cassette, several planes were simultaneously measurable.

Massiot (1974) clarified what happened in Lyon in 1935 which moved their interest away from tomography and towards planigraphy. The Messrs Massiot (Jean and his father) had exhibited a model of their tomographic equipment. The day after the presentation, J Massiot received a telephone call from Bocage, pointing out that his patent had not been cited. Massiot apologized that he had not seen the patent. A copy was sent and, shortly after, the Messrs Massiot decided to visit Ziedses des Plantes in Utrecht (see also §§ 1.3.2 and 1.4) and saw his experimental apparatus for planigraphy. The Massiot apparatus for planigraphy was designed as a copy of this equipment. The planigraphic equipment developed is shown in figure 1.11, taken from Naud's thesis. The first commercial section-imaging apparatus based on the principle of planigraphy was made by the Massiot company in 1936 and called the Planigraph (Littleton 1973).

Massiot's next equipment for planigraphy which used a circular movement was christened 'The Biotome of Dr Bocage' as a tribute to Bocage by proposing that he stand godfather to the apparatus. It was first demonstrated in October 1937 and was constructed from plans offered by Bocage himself (Bocage 1938, Massiot 1938). The Biotome was very successful and continued to be made until 1952 (Grigg 1965). The Biotome underwent extensive clinical testing in a factory in Courbevoie. Patients were taken to the factory through the winter of 1937–8 and the results were interpreted by Dr Kindberg, Chief of the Pneumologic Department in the Beaujon Hospital. Massiot (1974) in his brief history noted that considerable effort was expended to convince the experts of the time that circular blurring trajectories were to be preferred to others. In particular is mentioned Dr Maingot, Chief Radiologist of the Laënnec Hospital, in whose clinic there were two linear 'tomographs', namely a horizontal Sanitas machine and a vertical CGR machine both based on the Grossmann principle.

The hallmark of the Biotome was that the X-ray tube and film moved in vertical planes with the patient between. Typically the tube described a circle of some 40 cm diameter in about ½ s, giving a slice thickness of a couple of millimetres. Critics of the Biotome said that this movement might lead to errors and was overcomplicated (Bocage 1938) but on the contrary Bocage felt a spiral movement would have been even better. That this was not used was simply due to the increased technical complexity. Bocage described several experiments to verify the lack of error from a circular trajectory. It is in the paper by Bocage (1938) that there is confirmation that the date of his collaboration with the Messrs

Massiot was 1935, possibly dating from that telephone call after the exhibition at which J Massiot had failed to refer to Bocage's work.

The story was taken up by Buffé (1936) at a later meeting of the French Société. Buffé was a little more critical of planigraphy, noting that the degree of blurring of structures decreases towards the plane in focus. There is no clear demarcation between in-focus and blurred volumes— that is he was commenting on a partial volume effect. Buffé had available from Massiot equipment for both tomography (with linear pendulum movement) and spiral planigraphy. He much preferred the latter and had given up completely on tomography in this application. A word of nationalism appears in his paper: 'For many years Dutch, Italian and German radiologists have studied this method and applied it easily. Is it not long since French doctors, in their turn, made a place for it in their laboratories, notwithstanding that it was a French physician, Dr Bocage, who had the first idea?' Buffé's paper was commented upon by Arcelin who reminded Buffé of the first Strasbourg conference organized by Gunsett. Although there were over two hundred delegates, there were, he said, only three from Lyon and he named them. However, he now assured Buffé that his results at the military hospital Desgenettes were as good as those he had seen presented in Strasbourg. Incidentally Arceline claimed to be the first person ever to install and use three-phase supply in 1910 in the Hospital St Joseph in Lyon. He said his radiography was performed using glass-plate detectors, films then being rare. Screens were not commonly used and there were no Coolidge tubes! (see appendix 1).

Naud's thesis is a doctoral work divided into three parts. In the first he reviewed the historical developments. In the second he analysed the performance of several of these developments and in the third he devoted the latter half of the thesis to his own experimental data. Picking out a few key points from this work, we might note that he clarified how, even before Bocage, Mayer (1916) and Baese (1915–7, 1917a,b) were working along the right lines towards section imaging. We shall return to these two pioneers in §§ 1.2.12 and 1.2.13. This account is delayed until then because their work was not true section imaging although it nearly was! Naud also stated that Bocage began work in 1917. He is quite clear that Bocage did not build apparatus although some other reviewers would dispute this. (No doubt they are referring to the Biotome of Dr Bocage which was prepared from plans by Bocage but constructed by the Massiot company (Bocage 1938).) Naud's own work was performed with two machines, one due to Buffé employing linear motion and the other due to Massiot with spiral movements after Ziedses des Plantes. The work contains some beautiful examples of imaging pulmonary structure and structures in the cranium. The two principal products in the early years from the Massiot company, spiral tomography after Ziedses des Plantes and the Biotome of Dr Bocage, were described again by Massiot

(1938). The former essentially permitted the patient to be horizontal on a couch whilst the latter required the patient vertical between the vertical planes of traverse of the X-ray tube and film.

Body-section imaging was also reviewed by Morel-Kahn and Bernard (1935). Their review included details of most of the studies reported in this chapter. It is worth pointing out that the approximations of stratigraphy were well known to them. They referred to the work of Bistolfi (1934) whereby sections were sharply focused by rotating the body within the fan beam from a stationary source falling on a stationary detector (figure 1.32). The intention was for the plane parallel to the detector and passing through the axis to remain in focus, all other planes being blurred. What actually happens, of course, is that this is only true for the central point itself, as noted previously, and that all other points in the plane are blurred. This blurring becomes more pronounced as the distance from the point to the axis increases. So, for example, point A is 'fairly sharply imaged' but certainly does not image to a sharp point. Certainly the degree of blurring is far less than for a point not in this plane, such as B, even if B were closer radially to the axis than A. Morel-Kahn and Bernard were concerned that this stratigraphic technique led to 'déformations périphériques'. M Massiot (1935) also commented

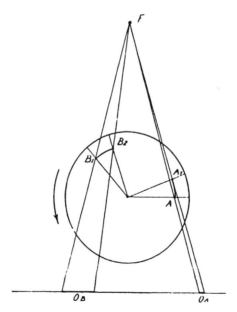

Figure 1.32 Bistolfi's apparatus for stratigraphy illustrating how the blurring O_A for a point A on the plane of interest is less than the blurring O_B of a point B off this plane. The object rotates whilst the source F and detector remain stationary. (From Morel-Kahn and Bernard (1935).)

on these inadequacies of stratigraphy, verified mathematically the tomographic technique of Grossmann and showed how a new planigraphic apparatus invented by his brother (J Massiot) also satisfied the equations for perfectly sharp plane imaging.

All this activity in France around 1936 was thus clearly set against a background of awareness of the many workers in the 1920s. Much effort was made to relate techniques and establish their chronology.

1.2.9 Derivative workers in planigraphy and tomography

We have now met most of the prime movers from the 1920s and early 1930s with contributions of fundamental importance to the development of the subject. Two others (Kieffer and Watson) are to come but before that let us take a sideways look at what was happening away from mainland Europe to capitalize on the inventions of the pioneers.

In the USA, Andrews and Stava (1937) also decided to construct their 'experimental tomograph' using the principles of tomography described in §1.2.6. The X-ray tube (or Röntgen tube as was the common description of the day) was mounted rigidly onto a pendulum B (figure 1.33) via an arm C. The cassette E was placed in a tray D which was kept horizontal by the system of levers G, the tray being pivoted at H. The plane of the film passed through the centre of the pivot H. The pendulum B was pivoted at P to a fixed vertical member, such that the position of P could be adjusted to the desired plane of cut. The cassette tray was positioned as close to the object as possible and locked horizontally at H and J. The

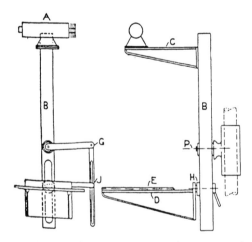

Figure 1.33 Andrews and Stava's apparatus for tomography; front and side views. (From Andrews and Stava (1937). Courtesy of the American Röntgen Ray Society.)

apparatus was started by rotating the pendulum manually to the required half-angle. A counterweight at the lower end of the pendulum B then caused this to oscillate when the pendulum was released. The firing of the X-ray tube was also easily controlled from switches activated by the passage of the pendulum. Andrews and Stava described experiments in which they had much success with the equipment (figure 1.34).

Figure 1.34 Photograph of the experimental tomograph of Andrews and Stava. (From Andrews and Stava (1937). Courtesy of the American Röntgen Ray Society.)

Stava (born 1906) joined Picker in 1934 as a mechanical engineer and stayed on until his death in 1953. Andrews (born 1906) (figure 1.35) was a US radiologist who made a detailed study of the European literature after moving from the University of Pennsylvania where he was a resident in radiology to Cleveland University Hospitals to work towards an MD. He first exhibited his equipment at a meeting of the American Röntgen Ray Society in September 1936. The 'new' device was reported in the 29 September copy of the *New York Times* but Brecher and Brecher (1969) record that most of the radiologists who stopped at the exhibit were unimpressed because, to the unaccustomed eye, tomograms initially appear to be more blurred rather than less so (for example the ribs in a chest exposure designed to show lung structure may not be visible—as intended of course!).

Chausse (1939) added a novel element to tomography. He proposed multiple firing of the X-ray tube along its track with unexposed intervals between so that there was no confusion between identifying a true line of structure from the plane in focus and that due to blurring a point not in focus into a line.

In the UK, Twining (1937a) was also experimenting with what he called 'tomography' at the same time. From his clear description of his apparatus (which follows the geometry of figure 1.4) he was, by the definitions of Andrews and Stava† (1937), actually performing 'planigraphy' and indeed had successfully devised a neat way of achieving what Andrews and Stava thought would be too difficult. Twining made use of the fact that the Potter–Bucky couch embodies lateral translation of the column carrying the X-ray tube and a film carriage moving along the long axis of the couch on its own rails. The tube and film move parallel to the couch and floor and parallel to each other at a fixed distance (as in figure 1.4). The option for the central ray to pass through the axis appears not to have been invoked. Twining's arrangement was to couple the two by a horizontally pivoted rod and fixed connecting rods (figure 1.36). By making the length of the pivoted rod AB equal to the focus-to-film distance and pivoting the rod at a distance AO from the film-connecting rod equal to the distance of the required cut above the film, the geometry described in figure 1.4 was achieved.

Figure 1.37 shows the quality of clinical tomography at this time. The radiation dose must have been tens of rads, given that intensifying screens were not in common use. The major advantage of removing unwanted shadows is apparent. This is one from an atlas of photographs based on the clinical work of Twining. Today's radiologists, used to the quality of

---

† Andrews and Stava (1937) and Twining (1937a) are not consistent in their use of terms. The former clearly define stratigraphy, planigraphy and tomography as reproduced here and summarized in the glossary and in table 1.3. Their experimental embodiment is clearly 'tomography' and yet they entitle their paper 'planigraphy'. Conversely Twining has built apparatus for 'planigraphy' and yet entitles his paper 'tomography'. In the course of describing it (last line on p 332) he also calls it 'stratigraphy'! Andrews (1936) goes some way to explaining the confusion. He prefers to regard *all* the techniques as 'planigraphy', reserving the use of the other two terms for specific embodiments. Here we prefer to retain all three original terms with the meanings as explained. Moore (1938) includes an extensive postscript on terminology alone. He did not like the term 'planigraphy', believing it to have the wrong associations with the printing industry or apparatus for changing the scale of drawings. He also thought that 'stratigraphy' had geological connotations whereas 'Tomograph' was Grossmann's trade name. The ICRU adoption in 1962 of the name 'tomography' for all body-section imaging is discussed in § 1.4.

Figure 1.35 J Robert Andrews, the first American to implement tomographic methods. (From Littleton (1976) *Tomography: Physical Principles and Clinical Applications* The Williams and Wilkins Company. Reproduced with permission.)

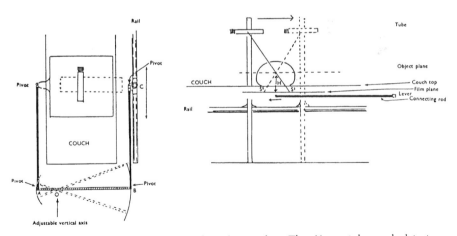

Figure 1.36 Twining's apparatus for planigraphy. The X-ray tube and detector mechanism were coupled via a pivoted rod AB itself jointed to connecting rods (shown dark). By varying the pivot point O the depth of cut could be selected. The arrangement was based on the standard moving tube and Bucky. (From Twining (1937a).)

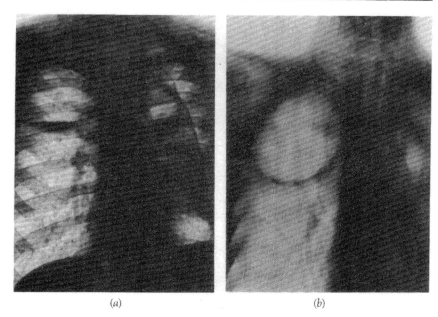

(a)   (b)

Figure 1.37 The improvement which tomography provided over planar imaging is illustrated by this clinical case. (*a*) Anteroposterior planar view of a patient with a cavity in the right upper lobe (containing fluid) and a cavity in the left upper lobe. (*b*) a Tomogram typical of the quality in the late 1930s showing the cavities clearly without obscuration by rib shadows. The fluid level is not seen because this patient was recumbent for tomography. (From McDougall (1940).)

computed tomography, would, however, be far from satisfied (see also figures in appendix 3 for examples of clinical tomograms).

Twining (figure 1.38) was a radiologist and the bulk of his paper concerns the clinical studies undertaken. Twining's eminence as a radiologist is today honoured by the Twining Medal of the Royal College of Radiologists awarded biannually to a Fellow of the College and first awarded in 1943. He was born in 1887 and died in 1939, not long after this work. Known as a man who combined innovative thought with manual dexterity he is remembered for many great achievements including a monumental text on radiography of the respiratory system, his study of the ventricles of the brain (the Twining line is named after him) and the work reported here on tomography. He was a colleague of Barclay and Patterson in Manchester. According to Burrows (1986) (where full biographical material is to be found) his mental energy possibly stemmed from a lifelong physical disability due to chronic osteomyelitis.

The first recipient of the Twining Medal was Professor Robert McWhirter who remembers Twining with some affection. McWhirter

Figure 1.38 Edward Wing Twining, Honorary Radiologist, Ancoats Hospital, Manchester, Assistant Radiological Officer, Royal Infirmary, Manchester. Photograph taken on the occasion of the Second International Congress of Radiology in Stockholm, 23–27 July 1928. (From Renander (1928).)

was appointed for a short period to the staff at Manchester Royal Infirmary in 1934 just as the equipment had been brought into use (McWhirter 1988). On Twining's practical abilities he wrote, 'I was invited to his home on several occasions . . . and was shown over his "workshop" which was equipped for metal work. He greatly enjoyed his hobby and had obviously become very accomplished. I can recall a lock he had made to fit a key—no mean feat and far beyond the capabilities of the hospital mechanic. Patterson worked in wood and between the two of them they could make almost anything.' The Twining planigraph became widely known. McWhirter says, 'I remember a foreign radiologist coming along one day and asking if he might be allowed to see the Twinograph! Needless to say this amused EWT greatly.'

Twining also considered a vertical lever, coupled to the cassette and tube either by pins in slots in the lever or by connecting rods. He rejected this idea on the grounds it would transmit too much vibration to the X-ray tube. Bush (1938), however, built such an apparatus for 'planigraphy' (figure 1.39) with the modification that all the movements were across rather than along the axis of the table. Strangely (more confusion!) he entitled his paper 'tomography' which it is not by the definitions of Andrews and Stava. This was good old-fashioned experimental physics! He built the prototype from broomsticks and Meccano

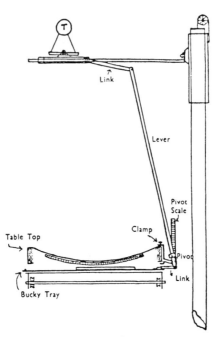

Figure 1.39 Bush's apparatus for planigraphy using a vertical lever. The Bucky tray travelled a small distance and the X-ray source a larger distance. The movement was transverse rather than longitudinal as it was in Twining's apparatus. (From Bush (1938).)

and operated it by hand. He claimed the method was 'provisionally covered by a patent' but a search at the UK patent office failed to locate a filed patent. It may be that his application eventually failed and this view seems to be supported in his short history by Bricker (1964). Bush described contacting a firm for manufacture and Kerley (1937) reported that the equipment was adapted by Messrs Newton and Wright. Bush was a little scathing of circular or spiral tomography, thinking they destroyed the smooth running. Indeed he thought the phrase 'what you may gain on the swing, you may lose on the roundabout' was literally true here. The need for an inexpensive tomograph despite Grossmann's developments was cited by many workers as the reason for Twining's development. Kerley (1937) for example pointed out how considerable space was required for the Grossmann equipment as well as this cost penalty. He reported clinical work done with a commercial version of the Twining equipment mainly in the field of pulmonary imaging.

There is some doubt concerning the exact validity of the mechanisms of Twining and Bush using pivoted connecting rods and a fixed-length lever. If the connecting rods had been rigidly fixed parallel in the planes of movement of the tube support and cassette and coupled via pins to *slots* in the pivoted lever, then the tube and cassette displacements would have been genuinely in the ratio of segments of the pivoted lever. However, when the coupling rods are allowed out of parallel and jointed to the lever (whose ends describe *arcs*) by pivots, the displacements are not exactly in the ratio of segments. This fact appears to have been overlooked, perhaps because the errors are not large for small displacements. Wheeler and Spencer (1940), also fired by reading Twining's paper, built an instrument for 'planigraphy' (this time the paper was titled correctly). Their embodiment used telescopic connections to the tube and cassette holders and thus *exactly* modelled the geometry of figure 1.4 with none of the difficulties mentioned above. At much the same time, Alexander (1938) also modified the Potter–Bucky table to achieve planigraphy (his paper is also entitled tomography). Unlike Twining, Bush, and Wheeler and Spencer who used some type of lever connecting the source and detector, Alexander used a system of pulleys (figure 1.40). By varying the number of pulleys and/or the source-to-detector distance, any desired plane could be kept in focus.

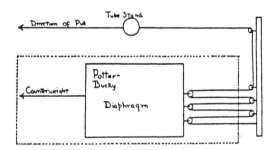

Figure 1.40 Alexander's pulley method for achieving planigraphy. (From Alexander (1938). Courtesy of the American Röntgen Ray Society.)

The system of Alexander was utilized clinically by Taylor (1938). Like most of the workers of the day he was interested in diseases of the pneumothorax and his paper contains many clinical examples which show the benefits of section imaging. Generally he used sections 1 in apart but in difficult circumstances the sections were closer. The opening of his clinical paper is a quite comprehensive historical list of the pioneers who had developed section imaging. In addition to the workers who have been covered in this chapter, he cites the work of Myron M Schwarzchild whom he said developed a planigraphic apparatus in 1931 and another in 1938. He references only a private communication and this worker does not feature in many of the other early papers. Taylor, however, cites his later apparatus as 'excellent for this type of work' and his name was linked with that of Twining as a pioneer of simple tomography by Cahill (1941). Watson (1939, 1940) reviewed many of the methods of actuating ratio movements and has also commented on the necessity for levers to be slotted to achieve variable length during the motion. He has drawn a modified version of Bush's equipment with the vertical lever slotted rather than utilizing fixed links. The same necessity was noted by Robin (1945).

In the USA, Zintheo (1939) was impressed by the technique developed by Twining. He referred to there being only two commercially available section-imaging devices in the USA in 1939. One was that based on Grossmann's principles and the other was developed by Kieffer (see § 1.2.10). Zintheo constructed his apparatus at Firland Sanatorium in Washington and reported in July 1938 that it had been in operation 'for some time'. It was a hand-operated machine with a horizontal lever. The tube travel was some 24 in with the beam on for half that distance. Exposures of 3 s were used. He had not yet automated the travel and method of switching on the X-rays which was necessary for shorter exposures. This paper was read at an Annual Meeting of the American Society of X-ray Technicians. In the audience was C J Bodle from Winnipeg who reported making similar developments. He said, 'These things are not very hard to make. We made one that cost us exactly nothing. It was all made out of junk from the scrap box (and) attached to the ordinary Bucky table.' Interestingly he commented that results were better if the tube travelled on a slight arc. These were indeed days of in-house development! Another radiologist with 'home-made' equipment in the late 1930s was Colyer (1937) (see § 2.2), and yet another to point out that planigraphy could be performed by adapting conventional apparatus was Bozzetti (1935).

The concern that, whilst tomography was of growing usefulness, particularly for diseases of the thorax, the Grossmann apparatus was prohibitively costly was widespread. In India, Viswanathan and Kesavaswamy (1940) wrote, 'We are of the opinion that for tomography a simple cheap contrivance can be designed to be attached to any serviceable X-ray couch, which provides for the free movements of the tube and the film carrier.' They constructed two such devices using pivoted rods attached to the Potter–Bucky and movable tube stand. One used a system of horizontal levers exactly like Twining's arrangement and the other had a horizontal rod pivoted to a fixed vertical rod to enable tomography in the vertical position. They also referred to an arrangement using pulleys like that of Alexander. The methods were obviously not very well known in the subcontinent because they concluded their paper, 'As it is not adequately recognized in India we hope that the publication of this paper will encourage the wider use of this method of radiographic studies by means of the simple and exceedingly cheap contrivance we have described.'

In fact, another 'home-made' arrangement had been reported in India the previous year. Two workers in Wanlesswadi constructed what they called a 'Tomoscope' using the Grossmann principle of tomography (Jones and Bradley-Bowron 1939). Their apparatus was similar to that of Bozzetti (see §1.2.6) in that they arranged for both the patient and the film cassette to rotate (coupled by the necessary lever) in front of a stationary X-ray tube (figure 1.41). They chose this implementation because the alternative with the patient stationary and the source and cassette executing a coupled motion necessitated a shock-proof tube which was unavailable to them. Their machine was delightfully simple, comprising two vertical axles (Ford front wheel axles) set 16 in apart in a solid wooden base. To one was attached a chair for the patient in the upright position (allowing, for example, visualization of fluid levels in pulmonary cavities) and to the other was attached the film cassette. The axles rotated through 90° during typically 1 s exposures. The cost of this equipment was stated at 50 rupees in contrast with the 1939 price of the Grossmann apparatus of 7000 rupees. Before construction a cardboard model was made to clarify the principles. The paper reporting this work shows several clinical examples of its use. It is noteworthy that significant

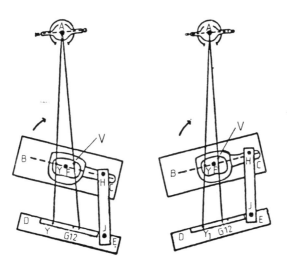

Figure 1.41 Diagrams (plan view) showing a cardboard model of the apparatus for tomography constructed by Jones and Bradley-Bowron. The equipment is shown in two orientations relative to the stationary X-ray tube A. The plane in focus is BYFHC and the upper part of the rotating assembly represents a view from above of a patient seated on a chair. The lever HJ couples the rotational movements. DE is the film cassette. The actual apparatus was constructed for tomography in the vertical position and blurring in the direction of the ribs (see text). (From Jones and Bradley-Bowron (1939).)

blurring of out-of-plane structures was achieved despite the fact that the apparatus rotated parallel to the ribs rather than orthogonal to the rib-cage as in the Grossmann machine. In chapter 2 are described further home-made apparatuses emerging from the need to improvise under War conditions (see §2.2) and even in the late 1950s some enterprising people were still improvising, for example Littleton in 1958 (Littleton 1976).

1.2.10 Kieffer and laminography

Added to the number of people who might rightly have been credited with inventing planigraphy should be the name of Mr Jean Kieffer of Norwich, Connecticut, USA (figure 1.42). Andrews (1936) has written as an addendum to his historical review paper (quoted verbatim by Moore (1939)) that he recently became aware of Kieffer's invention which he then felt took precedence over the claims of Vallebona, Ziedses des Plantes and Grossmann. The apparatus was invented in 1929 and a US patent was granted in 1934. Figure 1.43 shows a reproduction of the inventor's drawing. The principles of employing parallel fixed-separation motions (figure 1.4) govern the operation of the 'laminograph' (Moore

Figure 1.42 Jean Kieffer, French pioneer tomographer who introduced tomography to the USA. (From Littleton (1976) *Tomography: Physical Principles and Clinical Applications* The Williams and Wilkins Company. Reproduced with permission.)

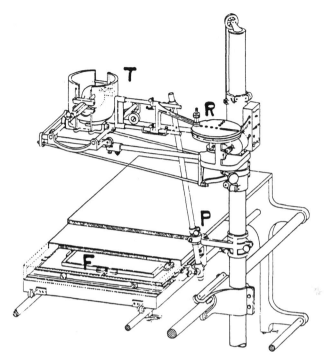

Figure 1.43 Schematic of the laminograph patented by Kieffer. (From Andrews (1936). Courtesy of the American Röntgen Ray Society.)

1938, 1939) in which compound motions of tube and cassette are possible including spirals, circles and linear shifts (figure 1.44). Kieffer's work was of paramount importance in establishing body-section imaging in the USA and the story of how it came about is intriguing.

Kieffer was a self-taught technician who himself developed tuberculosis and, after recovering, secured himself a position in a Connecticut tuberculosis sanatorium. It was there during a relapse in 1928 that he invented laminography in order to image the lesions in his own mediastinum. He told a reporter for the *Hartford Courant* (23 May 1937), 'I tried to get people to build a machine. . . . My proposals were met with sharp criticism. I went to X-ray manufacturers, only to be turned away with the assurance that the method probably wouldn't work. The head of one of the largest firms said, "It doesn't seem as if there could be anything to this idea of yours or our own engineer would have discovered it."' Brecher and Brecher (1969) tell how the idea came to see the light of day. Kieffer saw the headline in the *New York Times* (29 September 1936) 'New X-ray device "dissects" by films: machine makes possible photographs of parts of organs unobscured by tissue; takes slices of body'.

Figure 1.44 A commercial version of the Kieffer laminograph. (From Littleton (1976) *Tomography: Physical Principles and Clinical Applications* The Williams and Wilkins Company. Reproduced with permission.)

The article was a report of the machine built by Andrews and Stava which was being exhibited at the American Röntgen Ray Society September meeting (see §1.2.9). Kieffer at once set out for Cleveland. Brecher and Brecher (1969) write, 'Dr Andrews, who was exhibiting there a model of a machine which he called a "planigraph", clearly recalls Kieffer's arrival: "Towards the end of the 1936 meeting, an unassuming, unpretentious, and somewhat reticent chap presented himself to me while I was demonstrating our exhibit and told me, in almost these words, that he had been watching me for several days in order to 'size up' what kind of person I was and whether he would dare to approach me as he . . . considered himself the inventor of this apparatus, and it now appeared as if I were getting all the credit. This was all very strange, and so I agreed to meet him privately in his room in the hotel after dinner."' There Andrews told Kieffer the disappointing historical facts concerning the European patents. Kieffer must have been very dejected but in course of time things worked more in his favour. Sherwood Moore, Director of the Mallinckrodt Institute of Radiology, had stopped off at Andrews' exhibit and was impressed. Andrews initiated an introduction between Moore and Kieffer which in time led to Kieffer's ideas becoming a practical reality.

Much light is cast on the order of events by Kieffer (1938, 1939a,b) himself and it is worth quoting verbatim his statement. He wrote, 'I discovered the principles embodied in the machine independently (Kieffer 1929–34) in 1928, and designed a practical machine by 1929, being entirely unaware at the time of the previous work done by Bocage, and by Portes and Chausse in 1921 and 1922. On account of poor health, the then beginning Depression, and scepticism on the part of the men I tried to interest, I was unable to have the machine built until I met Dr Moore in September 1936. When shown the design and told of my hope that it would prove of clinical value, he immediately realized the possibilities of the machine and obtained permission from the authorities of Washington University to finance its building at the Mallinckrodt Institute.' Moore (1938) whilst clearly valuing his collaboration with Kieffer has recorded that there were difficulties associated with the fact that Kieffer, the inventor, lived 1400 miles from the centre where his device was being constructed and that subsequent clinical work had not been as rapid as it might otherwise have been. The apparatus construction was supervised by R Tontrup (Brecher and Brecher 1969). In his paper Kieffer went on to acknowledge the work of others on planigraphy but clearly felt they did not invalidate his exposition. Such personal writing would be very rare in a modern scientific paper but seemed common in the 1930s. Interestingly he is one of the few people to cite patents by Baese (1915–7), Bartelink (1931–2) and Grossmann (1934) in addition to those (much quoted) by Bocage (1921–2) and Portes and Chausse (1921–2). He does not appear to have been aware of the patents by Pohl (1927–32, 1930–2a,b).

Kieffer (1938) repeated at length the principle of planigraphy previously described, noting in particular how the character of the blurring and its amplitude depended upon the types of motion and their amplitudes and on the relative geometry of the levers connecting the source through the pivot to the detector. He stated clearly how there were two effects which planigraphy produced. The first, requiring a large amplitude of travel, was arranging for two erstwhile superposed structures to become unsuperposed for a significant part of the exposure. The second, having the opposite requirement of small travel, was for small-scale structure to be blurred by small amounts. He also stated the requirement for symmetrical blurring. The laminograph had sufficient flexibility of movement that the motions could be adjusted for the optimum arrangement for the object in question. Not only were the geometrical movements variable but also the speed of the movements was variable during the exposure; Kieffer (1938) provided a series of photographs illustrating these. He designed special templates to control the movements for more even blurring than could be obtained with an Archimedean spiral. All movements were electrically driven.

Kieffer went on to show (figure 1.45) how transverse axial sections

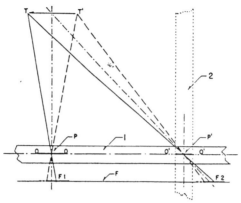

Figure 1.45 Kieffer's drawing explaining how the planigraphic principle could be extended to generate transverse axial sections. (From Kieffer (1938). Courtesy of the American Röntgen Ray Society.)

could be taken. In this arrangement the long axis of the body 2 is perpendicular to the planes of travel of the tube and detector and indeed these are on opposing sides of the body rather than above and below it. The same geometrical ratio arrangement holds whereby, as the tube moves from T to T' and the film is at F2, the section OO' remains in focus, other planes being blurred. This principle enabled, for example, sectional röntgenograms of the spine parallel to the vertebral discs. Even the requirement for the tube and film to move in the plane of the film could be dropped. Figure 1.46 shows how the transverse section OO' is

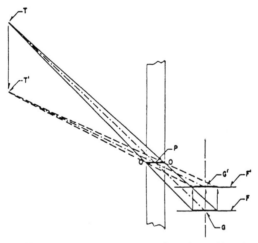

Figure 1.46 How a transverse section can be achieved by the planigraphic principle with the X-ray source and the film moving perpendicular to the plane of section. The source moves from T to T' and the film from F to F' and the plane OPO remains in focus throughout. (From Kieffer (1938). Courtesy of the American Röntgen Ray Society.)

still in focus when the tube moves normal to the film and the film moves normal to the film plane. What is more, there is no requirement for the film plane to remain horizontal; by tilting it, a sharp image of a plane with the same orientation in the body would result (figure 1.47). Figure 1.48 shows how Kieffer arranged for such configurations in practice.

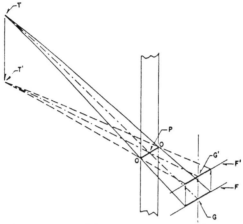

Figure 1.47 An extension of the principle shown in figure 1.46 whereby an oblique plane in the patient may be imaged by the planigraphic principle. The source moves from T to T' and the film from F to F' and the plane OPO remains in focus throughout. (From Kieffer (1938). Courtesy of the American Röntgen Ray Society.)

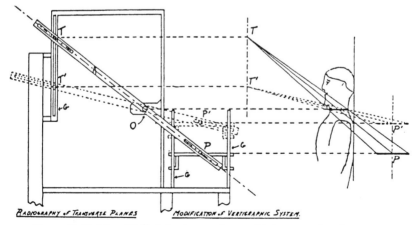

Figure 1.48 How Kieffer's laminograph was able to generate a transverse section by the method in figure 1.46. By tilting the film an oblique section would result as described by figure 1.47. In this figure the schematic representation of the body röntgenographed has been displaced to the right for clarity; in practice the points marked TT' and PP' on the right drawing are situated as in the left drawing. (From Kieffer (1938). Courtesy of the American Röntgen Ray Society.)

Once the general principle has been recognized that the only require-ment is for the relative *motions* of source and detector to remain parallel and in a fixed geometrical ratio, whatever is the actual *inclination* of the plane of the detector and the source, a large variety of possibilities arise. When the film and the X-ray source move in circular trajectories transaxial to the long axis of the body, so-called transverse axial tomography results (figure 1.49). Although the description 'transverse axial tomography' has been used for this method of Kieffer (Coulam *et al* 1981), whereby the source and detector execute circular motions, it is argued by Ziedses des Plantes (1988a) that a better description would be oblique tomography. Transverse axial tomography is strictly only achievable when the X-ray source makes a complete revolution *around* the patient, as invented by Watson (1937—9) (see §1.2.11). Kieffer's technique is no different from that of Bocage (1921—2) except for the orientation of the patient.

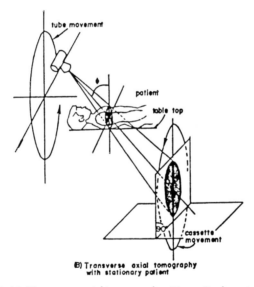

Figure 1.49 Transverse axial tomography. (From Coulam *et al* (1981).)

Kieffer recognized that his system was also suitable for röntgenoscopy if, instead of a film, a fluoroscopic screen was used. He described equip-ment to achieve this. He wrote, 'If the relation of the system to the body is changed during visual examination so that the selected plane is successively located through all planes of the body, the observer will see through the eyepiece each plane come into focus successively, and receive a visual impression similar to that received by a microscopist as he focuses through all the planes of a semitransparent solid illuminated by transmitted light.' Vallebona's wish, quoted in §1.2.1, seems to have come true. Indeed Kieffer recognized that his arrangement followed the laws of geometrical optics and that factors such as depth of focus, circle of confusion, etc, could be calculated. He said the laminograph was an 'X-ray-focusing machine' and that the system it used was a 'mechanical lens'. Note that in the fluoroscopic system the sharp focusing of one particular layer is a psychological effect of the fact that the eye can concentrate on those parts of the field which are unchanging whilst other structures move. There is of course no cumulative effect as would be the case with a photographic plate or film as detector. Incidentally Moore (1938), Kieffer's clinical collaborator, appears to have been one of the first people to point out that 'there is every reason to believe that preliminary röntgenography is essential and that therefore planigraphy will not supersede, and cannot be substituted for the standard preliminary examination'.

Also with reference to the work of Kieffer, he, like Portes and Chausse (1921—2) earlier, recognized the principles were also applicable to radiotherapy and made much the same kinds of statement. There is no evidence he himself followed up this idea. Figure 1.50 shows a diagrammatic representation of this idea by which radiation could be 'concentrated' at a small volume with the beam entering at multiple orientations, thus sparing dose to normal tissue.

Kieffer's (1929—34) patent comprehensively covers all these matters. A great deal of space is taken up describing the principle of planigraphy

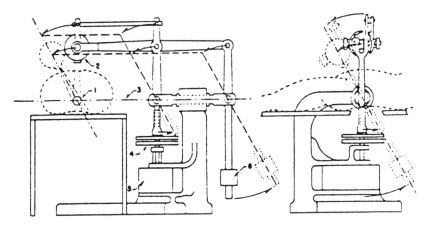

Figure 1.50 Kieffer's 'radiation-concentrating' device. The movement is along a section of a spherical surface. 1 is the tumour to be treated by the 'radium bomb' 2. 3 is the axis of the system and 4 is a cam and turntable for spiral motion. 5 is a motor and gearbox for driving the assembly. 6 is a counterweight. No matter where the system is displaced, radiation is always directed at the centre of the tumour. (From Kieffer (1938). Courtesy of the American Röntgen Ray Society.)

in both words and mathematical formulae including predicting the magnitude of blurring for out-of-focus planes in terms of geometry of tube travel and out-of-focus structure. One particular embodiment is described, again with much emphasis on pointing out that any equivalent embodiments are deemed to be covered by the patent. It is perhaps surprising that the examiners for this patent did not light on the earlier European patents since much of the discussion could equally well have been written about apparatus described in these. However, Kieffer sufficiently diversifies the number of ways he puts forward the principles that there seems little doubt he *thought* he was the first to do so. Picking out some interesting features in the patent, he refers to the radiotherapy 'X-ray accumulation effect' briefly in the introduction sparing 'injurious cumulative effects on the surface or in other parts of the body'. Indeed this is mentioned before the imaging application. He is also at pains to point out that the invention is not confined to medical use and that the use of medical terminology is not meant to imply this. Also he states that the principle applies to *any* penetrating radiation. He points out that, whilst the X-ray source plane and image plane remain equidistant, the source and shadow points for some particular object location vary in separation during the image formation and necessitate the use of sliding link rods at the ends of the pivoting arm (see § 1.2.9 where the necessity for this has already been discussed). He patented *any* general motion of tube and detector including both circular and linear planigraphy. In his patent he also pointed out that, whilst X-rays cannot be reflected nor refracted by ordinary lenses, the planigraphic principle was effective 'focusing' and that the laws of geometrical optics applied. He also paid a little attention to competing stereo viewing (see § 1.3.3) but thought this inferior to planigraphy. In summary, Jan Kieffer was certainly not the inventor of planigraphy but he surely made a substantial contribution to its historical development and his work led to the production of a working system for clinical use which popularized the method in the USA. Perhaps it is fitting (in relation to Bocage) that Kieffer (born 1897) was also a Frenchman. The Kieffer (first American commercial machine) appeared in 1938 and was called the Kele–Kieffer, manufactured by the Keleket X-ray company (Littleton 1973). This company, formed around 1903, by John Kelley and Albert Koett (Grigg 1965) was established in Covington, Kentucky, and gave to the town its nickname X-ray city.

Brecher and Brecher (1969) quote a personal letter from Kieffer to his wife, clearly showing his excitement at his success: 'I am so tickled at the machine and at what it does that I would like to dance a jig! Dr Moore and I almost did that yesterday morning when I got a film that he didn't think could be gotten . . . a picture of an abdominal aneurism! The impossible! . . . Boy oh boy! I was working on the machine when Dr Moore came up with the film as it came out of the dark room. "Come and see this," he literally yelled. ". . . It definitely showed things you wouldn't even guess with regular X-ray film. . . ." He was all smiles and shook my hand and said, "Kieffer, we got them all licked."'

1.2.11 Transverse axial tomography and Watson

In England in the late 1930s, William Watson (1895—1965) (figure 1.51), a largely self-taught radiographer, was also developing section imaging. His work spanned several decades and the importance of what he did has perhaps hitherto not been prominently recognized. As the story of section imaging unfolds here, we shall return to Watson many times.

Figure 1.51 British radiographer, William Watson, who among many achievements first described axial transverse tomography. (From Littleton (1976) *Tomography: Physical Principles and Clinical Applications* The Williams and Wilkins Company. Reproduced with permission.)

Quite a lot is known of Watson's early involvement in radiography in particular from a paper (Watson 1961) in which a deliberately personal account of early radiography was read to the Society of Radiographers. We might note that coincidentally Watson was born in the year X-rays were discovered. He was wounded at Ypres in 1916 after a 10 month spell of trench warfare, taken to the Canadian hospital at Etaples and 'skiagraphed'. This was Watson's introduction to X-rays. He was obliged to spend 2½ years as a patient subsequently at the Military Orthopaedic Hospital, Hammersmith (now the Royal Postgraduate Medical School). All patients were required to take part in a rehabilitation scheme and Watson chose to clean the brass in the X-ray department. He was soon promoted to the darkroom and began to study the theory of X-ray recording from photographic journals. He made 9p a day. In 1918, Captain Stanley Melville took charge of the Department and sent him to the Navy and Army School of Radiography at Imperial College. There he was taught physics by Major Phillips, a pioneer of X-ray technology who had been working as early as 1896 at the (now named) Royal Marsden Hospital. Watson played an active role in the Society of Radiographers early in the 1920s and contributed many articles to their journal describing his inventions. He wrote, 'It was in 1920 that I developed an idiosyncracy for trying to improve things.' That 'things' needed improvement is most obvious from his recollections in this paper of exposures which delivered surface doses of some 100 + rad and of radiographers who regularly took to knitting 'whilst the patient cooked; half a row of a sock for an ankle and up to two rows for a lumbar spine'. The improvement in X-ray detection during the 1920s including double-coated film and tungstate screens paved the way for the possibilities of tomography in reasonable times which Watson and others were able to achieve. Burrows (1986) has reprinted an account from Watson (1961) of what life was like as a radiographer in the early 1920s.

A paper published in 1946 (Jupe 1946) highlights the role of the physicist in the radiodiagnostic department of that time. Here we find the comment, 'A vast amount of work remains to be done on body-section radiography . . . and stereography.' Other jobs done by the 1946 physicist might come as a surprise to today's medical physicist, things like checking the currents supplied by the electric power company and checking electric instruments.

Indeed the appreciation of a physicist in the radiology department appears to have been the exception rather than the rule. Jupe wrote, 'Recently in the field of radiodiagnosis this important branch has tended to be regarded as unnecessary.' The pioneers of radiology were medical men who took photographs as curiosities and were assisted by a porter, dispenser or a clerk who then became an unofficial radiographer. Jupe observed, 'Another partner must be added and that partner is the physicist . . . it is necessary to realize that his outlook and method of approach differ entirely from those of the medical man. . . . The whole subject is built up on physics. . . .' His words were prophetic: 'What I think may well happen is that eventually every big hospital will have a department of physics, and that individual members of the physics staff will be allocated to various duties, some full time and some part time. One will, I hope, be allotted to the radiodiagnostic department of the hospital to work as part of the department's team. I think that at present there is need of a full-time physicist in the diagnostic department. This may be modified later when the more pressing problems have been worked out. It is essential that the physicist has access to a well equipped workshop with a trained mechanic.' Watson, though always referred to as a 'radiographer', was clearly such a 'physicist'.

Watson applied for a UK patent on 29 December 1937 (granted 29 June 1939) for apparatus to perform transverse axial tomography. Whilst this is a late patent in the history of body-section imaging, it is important since it led to equipment which achieved widespread use (Grover 1973) (see §2.4). It also represented several fundamental changes of thought concerning how the imaging should be performed. Firstly Watson (1937–9) recognized that radiotherapists would find *transverse sections*, rather than longitudinal sections, of great value in planning radiotherapy. He uses adjectives such as 'normal' to describe the longitudinal section, reflecting the contemporary practice of this method. Secondly he departed from the usual practice of moving the X-ray tube and detector relative to a stationary patient and introduced a revolution-ary concept. The usual stationary X-ray tube was utilized and the coordinated movement required by the planigraphic principle was of patient and detector. Watson stated that this was advantageous for tubes of delicate structure which may not withstand the shock of movement. Most of the patent concerns equipment based on these principles although he described briefly a second arrangement with the patient stationary.

Watson's apparatus may best be understood with reference to figure 1.52, from his patent. The film detector was placed at the top of a telescopic column which could rotate about its axis. Correspondingly the patient sat or stood on a rotating turntable 17, whose motion was coupled to that of the film by an endless chain 26 passing over sprockets 8, 18, 21 and 22. The X-ray tube was situated remote from these such as to cast an oblique shadow through a region of the patient encompassing the section of interest and onto the film. The section of interest could be selected by raising or lowering the film pedestal which was also provided with movement towards and away from the patient turntable, the slack on the drive chain being taken up by a pantographic arrangement of levers supporting the sprockets. The adjustable movements were locked whilst the study took place. During the exposure the patient and film could be moved through any angular arc including, if necessary, more than one complete revolution. The system was powered by an electric motor and Grover (1973) records that an experimental prototype was driven via the clutch and gearbox assembly of an Austin 7. Recalling the work of Jones and Bradley-Bowron (see §1.2.9) parts from motor cars seemed a popular cannibalization for constructing tomographic equipment.

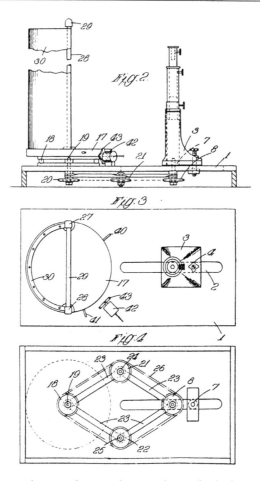

Figure 1.52 Schematic diagram showing the method of construction of Watson's patented equipment for transverse axial tomography. The upper drawing is a side view and the other two drawings are plans from above and below respectively. On the right is the apparatus for holding the detector whilst the patient is to the left. The X-ray source is not shown. See text for description. (From Watson (1937–9). Courtesy of the Comptroller of Her Majesty's Stationery Office.)

Watson's patent made no mention of the need to prove or demonstrate the mathematical principles of tomography and one might thus conclude that he considered these to be too well known (by 1937) to be worthy of further comment. There is, however, a most important additional purpose to his patent and that was to point out that, provided that the movement between patient and detector was precisely synchronized, there was no restriction of the film plane to the horizontal. The film could, for example, be vertical in which case the corresponding vertical plane through the patient containing the axis of rotation of the turntable would be imaged sharply. If this vertical film plane were offset a given distance from the axis of rotation of the carrier, then the sharply imaged vertical plane in the patient would be correspondingly offset. If the film were oblique, then an oblique plane in the patient would result. Watson drew film carriers for all these possibilities. Moreover, holders to receive a plurality of films were also described. Even weirder arrangements of cylindrically stacked and radial film sets were proposed. The same principles of generating sharply focused planes at arbitrary orientations had been described by Kieffer in the USA some years earlier (see § 1.2.10). Figure 1.53 shows clearly these principles for obtaining

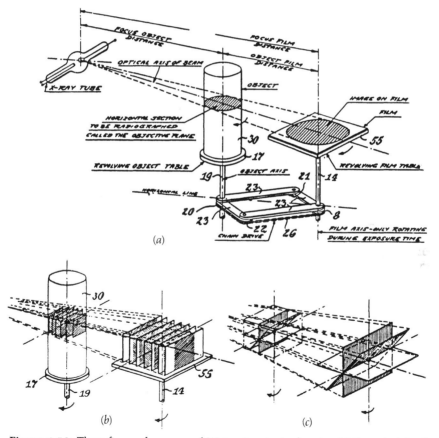

Figure 1.53 Three figures from one of Watson's patents showing (*a*) the principle of transverse-section tomography, (*b*) how several sections might be obtained simultaneously by using a book cassette of several films on the rotating turntable and (*c*) how oblique planes could be formed; the obliquity of the plane of section exactly matches that of the film on the rotating stand. (From Watson (1939–40).)

multiple sections at arbitrary orientations. Watson (1939—40) took out an American patent embodying very similar ideas to those in his UK patent.

It is interesting to note that the technique of obtaining multiple parallel sections by using book cassettes of many films was reinvented many times. For instance de Abreu (1948) describes his invention of this method and further work was reported by Watson (1950) and Lasser and Nowak (1956). The method has been described under several names, some rather whimsical (Grigg 1965), such as Multisette (for seven cuts) and Multisette Jr (for three to five cuts). Another commonly used term was poly-tomography and plesiosectional tomography. In passing we might also note that, in his paper, de Abreu claims to have demonstrated the principles of tomography in Brazil between 1924 and 1928 (see also de Abreu 1926, 1930). Manoel de Abreu was a Brazilian radiologist who had trained in Rio de Janeiro and Paris. He lived from 1892 to 1962. de Abreu is possibly better known for his invention of photofluorography which in Brazil is known as abreugraphia (Grigg 1965). He was Professor of Radiology at the Faculty of Medical Sciences in Rio de Janeiro and Director of Pulmonary Diagnostic Services in the city. He was also well known for philosophy and poetry and died of lung carcinoma on 30 January 1962 (Bruwer 1964, Grigg 1965).

Prior to the above-mentioned patent, Watson had taken out an earlier patent (Watson 1936—8). In this patent he described equipment which was able to make longitudinal body sections in addition to use in conventional mode for non-tomographic imaging. Essentially the idea is the one which he described in his 1939 patent as an addendum to transverse-section tomography. The patient was seated on a turntable and the detector was upright in a carriage which rotated in synchrony with the patient, thus defining a single vertical plane in the patient which remained in focus. His diagrams show the detector on the axis of rotation and thus the in-focus plane was that parallel plane passing through the axis of rotation of the patient. In this earlier equipment there was no provision for transverse axial imaging which was his later invention. He had, however, found that the blurring of other planes was more effective when the X-ray tube was not in a horizontal line with the centre of the detector and the centre of the image plane in the patient. By raising the X-ray tube vertically so that the beam was essentially downward directed, the blurring was more efficient. This idea was continued of course in the later patent.

The apparatus was described at some length in the patent and we shall not here comment in the same detail. The equipment is shown in figure 1.54. 30 is the detector, 17 is the patient turntable and the various rods, struts and bearings are to provide linear motions to a supporting frame such that the coupled rotational movement described above was achieved. The movements were all effected by spring pulls with damping

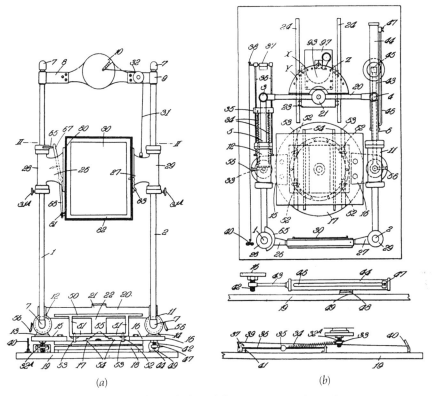

(a) (b)

Figure 1.54 The profile and plan of the apparatus in the earliest Watson patent. This apparatus was capable of generating both longitudinal sections and also stereo pairs, conventional planar projections and could perform fluoroscopy. See text for details. (From Watson (1936–8). Courtesy of the Comptroller of Her Majesty's Stationery Office.)

by fluid dashpots. Watson also explained how the equipment could be electrically driven. The details on the other side of the patient from the detector are the mechanisms for automatically switching the X-rays on and off. There was provision via lockable slide mounts for adjusting the height of the detector and also the proximity of the detector to the patient which it was desired to keep as small as possible.

The main achievement of Watson's apparatus was longitudinal-section imaging but he emphasized that his equipment was considerably more versatile than the special-purpose tomographs (presumably referring to the Grossmann Tomograph) which were *only* able to perform section imaging. Watson's equipment could be used in static mode for non-tomographic imaging. It was also capable of producing stereo pairs by taking films at just two angles and not exposing at intermediate angles.

It was also possible to replace the film detector by a fluorescent screen and to allow the doctor to make a sequential tomographic investigation of the patient without the need to record films at close spacing. Once the region of interest had been defined, selected planes could be recorded on film. He also explained how the apparatus, fitted with a fluoroscopic screen, could be used to localize foreign bodies by a method which dates back to Baese (1917a,b) (see § 1.2.12).

Watson's comprehensive knowledge of earlier equipment for body-section imaging is summarized in two papers he wrote for the Society of Radiographer's journal (Watson 1939, 1940). In these he systematically analyses the movements and methods of making movements in much of the equipment reported in this chapter. Additionally he describes a Pantograph (figure 1.55) in which once again one particular plane may be imaged sharply with others blurred. The X-ray tube and film move in parallel planes towards each other but such that the ratio of their distances from the plane of interest is kept constant through the motion. This is enough to guarantee the required objective.

Grover (1973) recollected that a transverse axial tomograph (later known as the Sectograph) based on Watson's design aroused considerable

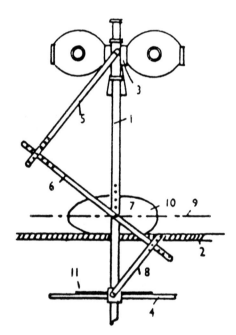

Figure 1.55 Watson's Pantograph for planigraphy. The X-ray tube 3 and detector 4 move towards each other along the rod 1, the distances they move being controlled by the lever 6 pivoted at 7 (which defines the sectional plane 9 in focus). (From Watson (1939).)

interest at the British Institute of Radiology Annual Congress in 1937 when it was viewed by Queen Mary at the Central Hall, Westminster. He also recalled that, as the gangway from the lecture theatre to the toilets passed the equipment, it resulted in 'phenomenal attendance at the stand'! Watson became a key figure in British radiological technology and further details of his life and work are in §2.4 including details of the commercial equipment resulting from Watson's invention of transverse axial tomography.

1.2.12 *The Baese patent*

It is one of the purposes of this chapter to draw together the varied contributions made towards the achievement of body-section radiography by many different workers in the early half of this century. Anyone looking at the literature from this time is drawn to the question of trying to decide to whom should go the credit for 'starting' the subject of section imaging. Some comment on this has appeared earlier including document- ing what the reviewers in the 1930s thought was the historical order of events. However, a desire to create a chronological sequence as the sole intention would be unhealthy and would camouflage the fact that the contributions made were of varying importance. For example a simple statement of how section imaging *might* be achieved could be construed of less weight than a description of an experimental embodiment of the principles which achieved clinical recognition. Also there is unfortunately no clear line between the time when there was no section imaging and the time when this was thought to be important, when the principles were worked out and equipment began to be constructed. However, for all this, Bocage appears to emerge as 'chronologically first' around 1921.

Or does he? What is the status of the patent filed 6 years earlier by Carlo Baese (1915–7), an engineer from Florence? The patent is entitled 'Method of, and apparatus for, the localization of foreign objects in, and the radiotherapeutic treatment of, the human body by X-rays'. In this patent there is a diagram (reproduced in figure 1.56) which contains the precise principle of planigraphy. There is another with the principle of stratigraphy. Imaging whole sections without blurring was not, however, Baese's concern. His interest lay in inventing a device which would yield the coordinates within the body of a foreign object such as a projectile. The principle was simple. He arranged for an X-ray tube T and fluorescent screen S to pivot about some axis C, the position of which could be varied in the body. The equipment was manoeuvrable relative to the body (or vice versa); that is the location of the pivot point was adjusted between studies and then kept fixed while the source and detector oscillated. When the image of the foreign body did not move on the screen as the source and detector oscillated, it was established that the pivot point coincided

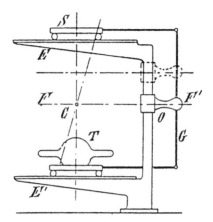

Figure 1.56 The historic figure in the patent of 1915–7 by Carlo Baese. It shows a method for localizing the depth of a projectile in the body (at C) by rocking a source T and detector S. When the projectile was at C, its image on the detector would be stationary with respect to this movement. This figure also of course can be interpreted as showing the planigraphic principle of body-section imaging. (From Baese (1915–7). Courtesy of the Comptroller of Her Majesty's Stationery Office.)

with the unknown location of the foreign body. Suitable scales were provided for reading the location. This is of course precisely the principle of stratigraphy. Indeed it is the *only* feature of stratigraphy that is mathematically correct, namely that the axis of rotation is the only line remaining unblurred as the equipment executes a rotation. Baese first gave suggestions for rotating the patient relative to fixed equipment (the stratigraphic principle). He then went on to explain that this was probably inconvenient and that it would be better to keep the patient fixed and to rotate the source and detector. It is in making this switch (figure 1.56) that he, probably without realizing it, 'invented planigraphy' for his diagram shows the source and detector moving *in parallel planes* via a pivoted lever arrangement. In this configuration of course, not only is the point C in focus but also *all* points in the plane FF' are in focus. Had he built this equipment he would be rightly the inventor of planigraphy. Unfortunately in the latter part of his patent, where he goes on to say that 'the constructive possibilities of the apparatus are infinite' and then provides a drawing of one embodiment (figure 1.57), he switches the principle back once again to having the source and detector rigidly fixed to the ends of a rotating pendulum. In this configuration, as we have seen earlier, only a line remains in focus rather than a complete plane, entirely adequate for his purpose but sadly falling short of the requirement for body-section imaging. It is interesting to note that Vallebona and Bistolfi (1935b) attributed the origins of their stratigraphic method to Baese. In the paper by Watson (1940) is drawn some practical equipment attributed

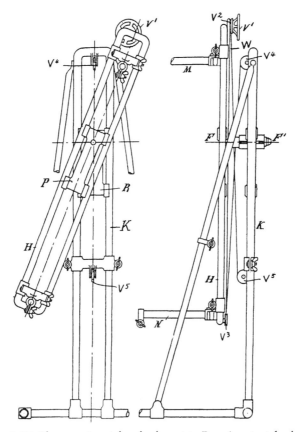

Figure 1.57 The experimental embodiment in Baese's patent for locating projectiles within a body. The equipment has departed from the planigraphic principle in figure 1.56 and instead shows apparatus based on a stratigraphic technique. A source M faces a detector N at opposite ends of pendulum H which pivots at F. Pulleys allow this pivotal position to be varied in the body. (From Baese (1915–7). Courtesy of the Comptroller of Her Majesty's Stationery Office.)

to the Baese patent for achieving movements in constant ratio using the planigraphic principle.

Baese was another worker to state that the imaging principle was also useful for radiotherapy in that the points of entry of a beam could be distributed whilst all being focused at a point. He is very particular, however, not to make any claim to have invented that idea. He said that at the time (1915) this was already well known.

Interestingly he also described the principle whereby a rocking mirror, whose movements were synchronized to those of the pendulum, viewing the screen enabled an apparently motionless image to be viewed. This is very similar to the idea put forward by Ponthus and Malvoisin (1937a,b) some 22 years later (see §1.2.8) and also exploited by Watson (1953b) (see §3.3).

Baese (1917a) wrote about his concern on several geometrical matters relating to ascertaining the depth of foreign bodies. He noted, for example, that a very small error in determining the lateral movement involved in shift radiography would lead to a large error in the depth measurement. For this reason he advocated taking a large number of shift films or the method of null movement embodied in his patent. The last paragraph of this paper is also an important pointer to the planigraphic principle. He wrote that radiostereometry had numerous applications like that of marking laterally on the patient all those points in the same plane as some foreign body which had been imaged, establishing essentially a cross-section.

There seems no doubt that Baese did not fully recognize the importance of his invention. Indeed Kieffer (1938) stated that the use of Baese's equipment to locate foreign bodies was a 'very limited application of the planigraphic principle'. However, there is a feeling amongst some patent examiners that, if a reasonable extrapolation from what is patented is obvious, then that extrapolation might also be thought to have been covered by the patent. In this sense what Baese wrote in 1915 was very significant in the history of body-section imaging. It is also interesting to note that, when Vallebona (1930d) described his stratigraphic equipment, he referred to the similarity between this and the method of rocking the film and tube used by Baese for the localization of projectiles in the First World War.

In 1918, Baese (1918) described his method again in English in a fairly extensive paper (which incidentally carried no references). He wrote that in the localization of projectiles there were so many solutions that it would be difficult to enumerate them all. However, radical reforms in equipment design were needed and 'we have endeavoured to free ourselves from everything which by force of habit had come to be looked upon as traditional'. From the text the new equipment appears to have become known as 'the Baese'. The stratigraphic principle was clearly described. Buried in the text is even the idea that a null-movement location method based on a circular trajectory of the X-ray source (that is Bocage's suggestion) would be useful. The ideas are all there!

1.2.13 *The work of Karol Mayer*

As early as 1914 the Polish radiologist Karol Mayer was working on techniques to reduce the confusion caused by unwanted shadows in planar radiographs. By choice his work is reviewed later rather than earlier in this chapter—which might at first seem strange in view of its date—because the significance of what he did was not really realized until

1935 when a remarkable exchange of correspondence occurred in the journal *Fortschritte auf dem Gebiete der Röntgenstrahlen*.

Mayer (1935a) had noticed the papers by Grossmann (1935b) and Chaoul (1935b) on tomography in the same journal in the previous volume. He wrote a single-page letter pointing out that the principles of tomography were grounded in the smudging of shadows by suitable movements of source and film. He noted that the origin of the technique was generally attributed to Bocage but said that he demonstrated a method based on the principles of shadow smudging at a congress of Polish specialists in Lemburg on 23 July 1914. The method was concerned with obtaining images of the heart without superposed structures. There does not appear to be a written record of this but the same work was described at length in a book (Mayer 1916) in Polish. The title of this historic book was *The Radiological Differential Diagnosis of Heart and Aorta*. Mayer stated that images of the heart were obtained by a method, the principles of which were not unlike those of tomography. He was prepared to admit, however, that the apparatus was somewhat primitive.

His method had an important limitation. The patient and the X-ray detector remained stationary and only the X-ray tube executed a movement. In drawings, this was indicated as a linear motion which could be either longitudinal or transverse to the body axis. The body being radiographed was placed very close to the film. By this means, fairly sharp images of the regions nearest the film were obtained whilst the images from the other parts of the body were blurred or smudged. The method was also used to obtain images of the head.

Hence Mayer (1935a) laid claim to have invented the principles of smudging by movement although he acknowledged what he described as noteworthy and ingenious subsequent inventions. His remarks were commented on by Grossmann (1935d) who agreed that Mayer had aimed at achieving blurring through motion but considered that he had not had the intention of creating body-section images. Grossmann felt that Mayer's book had not indicated the need for the film to move and that this idea was solely the intellectual property of Bocage. Mayer (1935b) replied to Grossmann's remarks. He asked permission to give an explanation of the claim which he had made and repeated that he had given the first concrete solution to the problem of shadow blurring. He said that his priority for this was public knowledge. His claim was based exclusively on work at these frontiers and he made it clear that he had no claim to the concept of film motion which he said was 'undoubtedly' due to Bocage. Further correspondence occurred between Vallebona and Bistolfi (1935a) and Mayer (1935c) on much the same lines.

Mayer's position in the history of Polish radiology has been elucidated by Grigg (1965). After the First World War the then separate Polish provinces were reunited to form independent Poland and Mayer took the first Chair of Radiology at Poznan in 1923. When the national radiological society was founded in 1925, Mayer became its first president and subsequently he was re-elected many times. He was born in 1882 and died in 1946.

Important detail from Mayer's book has been translated by the Polish radiologist Blaszczyk (1988). Mayer began his book by saying that, in view of the large number of organs in the mediastinum, the clinical assessment, for example by palpation of this area, was very difficult and he stated that in the 'pre-radiologic era' the diagnosis of retro-sternal thyroid goitre or cardiac diseases in patients with emphysema was hampered by the superimposition of shadows from other organs. Assessment of the heart was particularly difficult and it was to this problem that Mayer addressed his method. Consolidations in the lung representing inflammatory or malignant processes usually obliterate the contours of the heart. To overcome this Mayer wrote, 'After the routine preparations to the taking of the radiogram, I make fast but small (up to 8 cm) movements with the X-ray tube to and fro in the longitudinal or transverse axis.' He said this was technically straightforward because the X-ray tube is mounted in a movable box. No special or custom-made equipment was used, simply the standard radiology set-up.

Mayer apparently got the idea from observing an inkwell through the lid of a dustbin which was like a mesh. When he held the mesh still, he could not make out the outline of the inkwell but, when he moved it, the inkwell was clearly visible. He felt there was an obvious radiologic analogy. Essentially the distance which a structure moves on a film is proportional to the distance of that structure from the film. Hence it was logical to arrange for the heart to be as close to the film as possible. He recognized that, the smaller was the volume of pathology causing the unwanted shadows and the further this was from the film, the easier it was to get rid of the shadows. Even if the shadows cannot be entirely removed, they can at least be separated from those of the heart. Mayer tested his method on normal subjects from the medical clinic of the Jagiellonian University and concluded the technique was very satisfactory. His book gives many beautiful examples of the work. There is a strong similarity between the ideas expressed by Mayer and suggestions which came from many workers later that sharper pictures could be obtained by asking the patient to move certain structures (such as the jaw directly or the rib-cage by natural breathing) during routine radiology with a stationary tube. This had the effect of blurring these structures. The same principle is the basis of pantomography and both these are reviewed in §3.2.

1.2.14 The Vieten patent

Before we conclude looking at the earliest work in body-section imaging, it is worth noting two rather odd but nevertheless interesting radiological techniques. Heinz Vieten (1936–9) took out a patent for an ingenious, if slightly impractical, method of obtaining a transverse axial tomogram. The method is shown in figure 1.58. Vieten proposed to make a series of very narrow band-shaped tomograms at gradually increasing heights in the patient, the difference of the levels being equal to the widths of the tomograms. The moving detector consisted of a film and lead screen combination. The lead screen contained a slit which is the same size as the collimated X-ray beam and which defines the width of the tomograms. Between the tomographic exposures the film is sequentially shifted behind the slit.

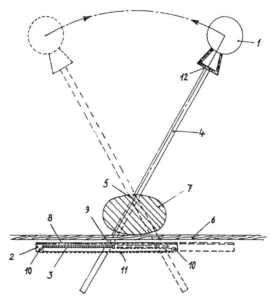

Figure 1.58 Vieten's method for obtaining a transverse axial tomogram. The X-ray beam 4 is narrowly collimated in the plane of the diagram and has a wide extent normal to this. The X-ray tube 1 moves as shown during the exposure of a small region 5 of the body 7. A film 3 with a slit collimator 8 moves in the contralateral direction. On each pass a strip of film only is exposed corresponding to the parallel strip in the body through 5. Sequential exposures are made with the pivot point successively lowered in the body and the collimator slit moved to sequential positions. The result is a tomogram *perpendicular* to the bed 6. (From Vieten (1936–9).)

The resulting picture of the joined band-shaped tomograms was a transverse tomogram *perpendicular* to the table top (Gebauer 1956). The procedure did not become popular and even Vieten himself was critical of the method (Ziedses des Plantes 1988a). Heinz Vieten (born 1915) was a

German radiologist. His concept is a connecting link between old and modern imaging techniques because it was a very early method of generating a *transverse axial* tomogram.

1.2.15 Heckmann's method

A form of radiology which at one time was most popular was that using so-called slit methods (reviewed by Vuorinen (1959)). A narrow slit or series of slits collimated the beam in direct contrast with area radiology. One particular slit method due to Heckmann (1939) bears an interesting relationship to tomography. Indeed it has been described as reversed tomography. The method is shown in figure 1.59. Once again the X-ray source moves from position I to II relative to a stationary film and body. However, a slit at a_2 moves proportionally along a linear route so that at any one time the object is only bathed in a narrow band of radiation. The movement of the slit and tube is such that the line between them appears to pivot about a point (shown in the figure as where the two shaded areas meet) which need not necessarily be in the body itself. It may now be appreciated that those areas of the body closest to the film are recorded as the sharpest whilst those near the pivot point are highly blurred. It is in

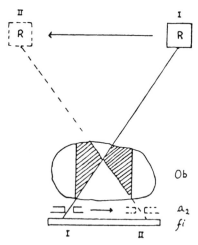

Figure 1.59 One of Heckmann's methods for obtaining a slice with little blurring. The tube R and the slit a_2 move along a linear route in opposite directions. (The tube moves from I to II.) The angle of incidence of the rays having access to the film varies during the whole exposure. As a result of this, the shaded area of the cross-section of the patient Ob is unirradiated. The planes of the body nearest the film fi are maximally sharp whilst the point where the shaded regions meet is maximally blurred. (From Vuorinen (1959).)

this sense that the method is reverse tomography in that generally (for tomography) we think of the plane containing the pivot as the sharpest rather than the least sharp plane. The method is in a sense also an extension of Mayer's moving-tube method (see §1.2.13) in that the region most in focus is nearest the film and other regions are blurred. What distinguishes Heckmann's technique is that there are volumes of the patient which receive no radiation at all (shaded) and in some circumstances this can be advantageous.

1.2.16 Planigraphy and tomography; physics of performance

Many of the early papers on planigraphy and tomography were remarkably lacking in any serious evaluation of physical performance, whether by experimental or by mathematical means. Generally workers concentrated on careful descriptions of the engineering aspects of the equipment developed and supplemented these with clinical examples of their use. If the techniques were being invented today, the scientific community would demand intercomparative evaluations and performance specifications. There are plenty of modern papers making up for these deficiencies, with retrospective analyses of the historical equipment and the modern counterparts which it has spawned. A particularly comprehensive account has been written by Coulam *et al* (1981). In the context of the current book it would not be appropriate to go into elaborate details but some aspects of performance will be mentioned, particularly as this casts further light on important differences between planigraphy and tomography and subcategories of these. By the middle to the end of the 1930s the papers were more analytic in their approach. Kieffer and Grossmann, for example, published detailed analyses of the physics of performance and related the quality of the images obtained to the controlling physical factors. Careful experimental performance measurements were more common once commercially manufactured equipment was widely in use (see, for example, Edholm 1960).

For planigraphy with a linear blurring movement, the plane in focus is selected by adjusting the fulcrum position on the arm linking source and detector. As the fulcrum is moved closer to the source, the angular movement increases for a fixed linear travel. Consequently the section thickness decreases and the smallest sectional thickness is achieved with the widest arc swing. At the same time, image magnification increases. The distance of the source from the detector varies during the exposure in linear planigraphy and hence there is a non-uniform X-ray intensity at the detector.

Planigraphy with complex blurring movements achieves not only a more complete blurring but also a more complicated sectional thickness, the effective thickness being determined by the time spent in moving from the largest to the smallest angle of inclination of source to detector normal.

For tomography, the plane in focus is selected by adjusting the height of the table and supporting the patient relative to a fixed fulcrum; the tomographic arc of swing remains the same for all sections which thus have the same thickness and the same magnification.

For both planigraphic and tomographic systems the image unsharpness in the focal plane increases with increasing arc of swing. There thus arises a trade-off between section thickness and unsharpness. The spatial resolution is also affected by the X-ray focal spot size and the type of intensifying screen used (in a modern system). Image distortion is also a variable function of position in the focal plane for planigraphic systems (because of variable magnification) and is fixed for tomographic systems. A particularly important (though less easy to quantify feature—since it depends on the subject being imaged) is the degree of non-focal plane information which appears to be in focus in view of the character of the motion used. The human observer will naturally view such an artefact as lying in the in-focus plane. Generally these effects have been evaluated experimentally.

Scattering of photons in the patient is also a contributing factor in determining the contrast which can be achieved for 'in-focus' planes and different imaging geometries have associated with them differing scattered contributions to the images.

Westra (1966) has published a particularly comprehensive book on the physics of tomography including discussion of 'zonography' which is the name often given to tomography with small angle of swing or with small blurring. Although some years old, this probably remains the best source of information for those wishing to pursue this subject in depth. Another modern account, by two of the pioneers of tomography, of the physics of tomographic imaging has been written by Vallebona and Bistolfi (1973).

1.2.17 Summary of the pioneers

The passage of time may have blurred the distinctions between the different techniques of the pioneers of the 1920s and 1930s in a way they would surely have disapproved of. The rivalry was fierce and no sooner was one claim for priority published than another rebuffed it. Perhaps in present times when lines of scientific communication are well established on an international scale it is hard to see how such important inventions remained unbroadcast for so long. The European origins of section imaging with patents by Bocage, Portes, Chausse, Grossmann, Pohl, Ziedses des Plantes and pioneering work by Vallebona, Bartelink, Ponthus, Malvoisin and the Messrs Massiot did not filter through to the UK and the USA until around 1935. In mainland Europe there would seem to have been a remarkable lack of interest amongst manufacturers in the early days. When equipment did finally become available commercially, it was necessarily expensive and many derivative workers took up the challenge of constructing simple copies for their own use. In the UK, Twining and McDougall enthusiastically embraced the methods and, in the USA, Andrews and Kieffer stand out as prime movers. The roles of Baese and Mayer are intriguing.

Most of the pioneers were lone workers and almost none of them would describe himself as a physicist. Yet the equipment they designed, patented and built had received careful thought and was elegantly conceived. Claims for priority in science are often acrimonious though, if one uses the dates from patents as the basis for sorting this out, the ranking amongst the early workers is quite clear with Bocage, Portes, Chausse, Pohl, Bartelink, Grossmann, Vieten and Watson lining up in that order. However, such a mere list ignores the fundamental importance of Ziedses des Plantes and Vallebona. The reader can make up his own mind where he feels the weight of credit should lie but I hope in telling the tale my own heroes stand out.

1.3 Multiple-film slice focusing, three-dimensional viewing and measurement

All of the above techniques for creating a sharp image of some particular slice of a patient with other slices blurred generate a single film for each study. At much the same time as these developments were taking place, other workers put forward methods by which specific slices could be brought into focus by special geometrical arrangements of a *number* of films, taken with the X-ray source and detector at different fixed geometrical locations relative to the patient. Between movements no exposure was made. We now consider such arrangements and will see that some of the prime movers of body-section imaging appear again.

1.3.1 Planeography and Kaufman

Kaufman (1936a,b, 1938) invented a technique which he called 'planeography'. The technique was elegantly simple. With reference to figure 1.60, the method was as follows. An X-ray source was moved from position T1 to T2 at a fixed height H above a plane in which shadows could be projected. Consider any two points A and B in some plane parallel to the shadow plane and at a height h above it. It follows from simple geometry that the projections of the line AB from the two source points T1 and T2 are parallel. The spread between their two shadow locations is also parallel to the line of movement T1T2 of the source and

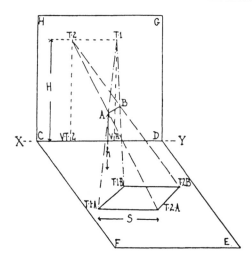

Figure 1.60 A schematic diagram showing the principle of forming multiple exposures for use in planeography, serioscopy or seriescopy. See text for explanation. (From Kaufman (1936).)

is simply dependent on this distance and the height h of the line AB. Kaufman decided to call the graphs which yield the spread from these two distances 'standard depth curves'. For a known separation between the X-ray tube locations, the spread could be read off as a function of the height h. Kaufman proposed that, instead of placing a single film in the shadow plane and exposing it twice, two films should be identically placed sequentially for the two exposures. When these two films were placed on top of each other above a viewing box, selected planes could be brought into focus by choosing to offset them laterally by the appropriate spread for some plane height h and tube separation. Information from that plane superposes in the composite view whereas shadows from other planes are blurred. By sliding the two films slowly across each other, a 'walk through' the object was possible. Kaufman thought this was the most useful way of using the films although he also explained how the two films could be used to make quantitative measurements of distances and angles. He considered that the greatest use for the planeographic technique would be in assessing asymmetries, rotations and involvement of structures by pathologic lesions. Drummond and Schmela (1939) extended this work to provide tables of direct use for making measurements. Kaufman made the statement that 'a röntgeno-gram may be considered the integration of serial planeograms'. Whilst this is obvious it was in a sense one of the earliest statements that a projection X-ray image can be related to the sum of corresponding tomographic slices.

90

1.3.2 Serioscopy; Ziedses des Plantes, Cottenot, the Messrs Massiot and Bocage

Cottenot (1938) published a technique which he called 'serioscopy'. In this paper he stated that the origin of the method of serioscopy was with Ziedses des Plantes (1936). The principles are similar to those of Kaufman's 'planeography' and it would be reasonable to regard these as synonymous. In Ziedses des Plantes' and Cottenot's technique, four radiographs are taken with the tube at four different locations. Once again the plane of interest is brought into focus by superposing all four on a viewing box and sliding them by the appropriate shifts. Apparatus for arranging this is shown in figure 1.61. Cottenot proposed that a vertical tube with embedded numbers be imaged simultaneous with the object exposure. The device acts like that of Ziedses des Plantes described in § 1.2.3, the numeral in focus giving the distance of the selected slice from some plane. Cottenot proposed that the four radiographs should comprise two pairs taken orthogonally for successful blurring. He was interested in applying the technique to imaging the thorax but realized the necessity for all four images to be exposed at the same moment of inspiration and designed a neat method of achieving this. The details need not concern us here except to note that the basis was a pneumatic belt worn by the patient and that the equipment was adjustable so that the serioscopy could be carried out at any part of the breathing cycle. Cottenot used a 'serioscope', a powerful viewing box in which the four films were placed

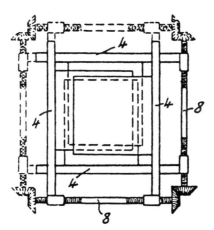

Figure 1.61 The seriescope of Ziedses des Plantes. Four films could be attached to holders 4 and thus arranged to lie over each other. The films were allowed to slide relative to each other by a mechanism whereby rotation of any one of the spindles 8 caused all the others to rotate (via meshed gears) and transfer this to a linear movement of the films via tapped bushes attached to the film holders. (From Ziedses des Plantes (1936). Courtesy of the Comptroller of Her Majesty's Stationery Office.)

in carriers which were synchronously coupled so that by turning a single screw the four films slide in unison, selectively focusing planes of known location (figure 1.61).

Kaufman and Koster (1940) noted the work of Cottenot and clearly stated their view that this was a derivative of their own method. In fact, as we shall shortly see, the method was invented in 1936 by Ziedses des Plantes and passed to the Messrs Massiot in Utrecht and thence to Cottenot. Kaufman and Koster wrote, 'The basic principles employed in serioscopy by Dr Cottenot are the same as those elaborated by us under the term "planeography". Apparently Dr Cottenot rediscovered plane-ography independently terming the method "serioscopy".' They gave the extra requirement that the axes of tube shifts not only must be perpendicular to each other but also must bisect each other for serioscopy to be mathematically exact. Kaufman and Koster, however, are critical of serioscopy, believing that, in view of the fact that the serioscope does not correct the enlargement of each planeogram or the displacement, an erroneous impression of shape, size and relative positions of structure would be obtained by the viewer. Kaufman (1936) had previously explained how distances measured within the planeogram required to be scaled to obtain true distances in the patient, whereas angular measurements did not require this scaling. Kaufman and Koster also set their technique in relation to those described in §1.2. They write, 'The term planeography was coined by us to differentiate this method from planigraphy, stratigraphy, etc. Regarding this term Jean Kieffer in a personal communication to us wrote as follows: "I think your differentiation of names is a very good one. Although the basic mathematics are similar to planigraphy, the procedures and results are different."'

The interest in France in the mid-1930s in body-section imaging was reviewed in §1.2.8. In particular it was noted that the Messrs Massiot had visited Ziedses des Plantes (in February 1936) and had compared results from his planigraphic equipment with tomographic apparatus developed by themselves. This visit provides the connection between the work of Ziedses des Plantes and Cottenot. Massiot (1937) reported that, on this visit to Utrecht, he had also seen a Seriescope developed by Ziedses des Plantes (1936). This equipment was a prototype of that described by Cottenot, namely four films, taken with the X-ray tube at four different equispaced stationary locations on the circular path of the X-ray tube were superposed in the Seriescope. By causing them to move relative to each other a sectional tour through the volume of interest was made. Ziedses des Plantes requested Massiot to copy this equipment. This done, a copy reached Cottenot.

Ziedses des Plantes (1936) had patented the technique and called it seriescopy (note the 'e' instead of the 'o'). In this patent, not only is the technique of moving films reciprocally to give the observer an impression of 'moving through the object' described whereby films are superposed on a viewing screen, but also optical methods for achieving the same results are proposed. By these the films need not be superposed but the images of all the contributing films are brought together by systems of mirrors so that the observer appears to be viewing a single film. By rocking the mirrors—which effectively leads to a translation of images with respect to each other—the same 'walk-through' effect can be engineered. In all seriescopy the density of the contributing films has to be reduced in inverse proportion to the number of such films. It is in this patent that the device for automatically recording the level of the plane in focus was described.

We saw in § 1.2.8 how Massiot had, on exhibiting tomographic apparatus, been contacted by Bocage who announced his earlier patent. After visiting Ziedses des Plantes in Utrecht, the younger Massiot informed Bocage of Ziedses des Plantes' invention of seriescopy. By a curious coincidence Bocage had described similar ideas to the elder Massiot only the day before. He had therefore to be told that Ziedses des Plantes had beaten him to it and supporting evidence of a paper from Ziedses des Plantes was produced.

Massiot (1937) makes the relationship between seriescopy and planigraphy quite clear. Seriescopy in no way substituted for planigraphy. It was done to guide the radiologist towards the precise section of interest which was then captured sharply by planigraphy. In questions, Ponthus criticized this technique as rather lengthy. She also criticized the choice of the word 'seriescopy' which implied radioscopy rather than radiology. Massiot disagreed, saying that, since the word was coined by Ziedses des Plantes, who were they to judge it a poor choice? Massiot refers to the joint work with Cottenot. Ziedses des Plantes' (1988a) recent comments on the sequence of events are in § 1.4.

Bush (1939) wrote in the *British Journal of Radiology* that in the UK 'the method known as seriescopy has not, I think, received quite the attention it should have in this country'. He noted how the method having been invented by Ziedses des Plantes was used by him largely for investigating the bony skeleton and that Cottenot in Paris had extended the method for thoracic investigations. Bush visited Cottenot in Paris in 1938 and developed a very similar way of arranging for the X-ray tube to be fired at precisely the same phase of inspiration for lung imaging. He took delivery of a Seriescope from Massiot and found the method of great clinical utility. Some of his paper is freely translated from Massiot's publications (and acknowledged) but this is another interesting example of a paper which clearly built on the work of others but which contains not a single reference. This paper to the British Institute in April 1939 emphasized Bush's view that seriescopy should precede planigraphy (to define regions of interest) and not replace it. The many careful practical matters to be attended to for successful seriescopy (such as patient immobilization, precise and equal tube shifts and care to reduce film exposure) were discussed in some detail. The first application in his list was the localization of foreign bodies, reflecting the current hostilities. A noteworthy quotation from his paper might apply to the whole subject of body-section imaging: 'Most new inventions are but the result of certain principles, already well known but perhaps forgotten, being brought together and reapplied in a new way by an imaginative mind.'

1.3.3 Stereo imaging

The origins of stereo imaging have been attributed to Elihu Thomson (1853–1937) (Grigg 1965, Brecher and Brecher 1969). Thomson was a physicist, founder and director of General Electric's first research laboratory at Lynn, Massachusetts. He was born in Manchester and came to the USA as a boy in 1858. He was a founder of the Thomas–Houston Electrical Company which merged in 1892 with the Edison Electric Company to form the General Electric Company. He first described X-ray stereo imaging on 11 March 1896 (Thomson 1896) explaining how he made stereo images of phantoms comprising metal objects buried in cork and wood as well as images of a mouse. It was also in this paper that he described the possibility of stereoscopy using a Crookes tube with two spatially separated cathodes. A short biography appeared in Bruwer (1964).

Bruwer also documented how an almost contemporary (10 March) development of X-ray stereo imaging was reported from France by Imbert and Bertin-Sans, two professors of medical physics in the faculty of Montpellier (Imbert and Bertin-Sans 1896). Professor Czermak at the University of Graz (Czermak 1896) also reported early stereo imaging. Apparatus was described whereby stereo images could be captured on a single plate and experiments were performed imaging salamanders. Figure 1.62 shows one of the earliest recorded stereoradiographs. A history of stereo imaging, which became quite popular (see, for example, Wilson 1911, Watson and Sons Ltd 1912), was compiled by Keats (1964). Haenisch (1911) writing his 'Röntgenological impressions of a journey in the United States' reported, 'The stereoscopic method . . . is widely used. In no single instance did I see a private laboratory which had not a more or less efficient apparatus for the making and the examination of stereoscopic plates.' The literature rapidly exploded with new techniques for stereoscopic imaging (see, for example, ARR 1913, Holland, 1919, Hill and Barnard 1919, Burdon 1919) and atlases of images (see, for example, Case 1916, Jaches *et al* 1916, Dunham 1916).

The use of stereo pairs or stereoröntgenometry for measuring distances within solid objects is very old, greatly pre-dating all of the developments

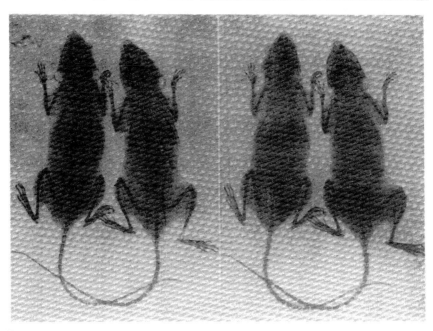

Figure 1.62 One of the earliest stereo pairs. (From Czermak (1896), cited in Bruwer (1964) *Classic Descriptions in Diagnostic Radiology* vol 2. Courtesy of Charles C Thomas, Publisher, Springfield, Illinois.)

with which we have been concerned so far. The reason for not reviewing it earlier is that stereo viewing has somewhat different objectives, namely to yield an apparent three-dimensional view of an object rather than the aim of isolating sharply some particular plane in the object. The subject is, however, closely related to body-section radiography using similar geometrical relationships and is here covered briefly.

The basic principles of viewing and measuring (Mackenzie Davidson 1898, 1916) go back to Sir James Mackenzie Davidson (1857–1919), a former ophthalmic surgeon from Aberdeen who was knighted for services to radiology in 1912 (ARR 1913). The cross-thread method of locating a foreign body such as a bullet or for making measurements of distances between points in tissue originated in the last century. Sweet (1903) described having used the method for at least 6 years. The technique was developed by Manges (1911) (born 1876) for measuring the bony pelves of pregnant women. Figure 1.63 which is taken from Johnson (1935) shows the method. X and Y represent extremities of the distance to be measured. Two exposures are made with the X-ray source sequentially at A and B casting images BX, AX, BY and AY. By placing the two images sequentially in a 'stereoröntgenometer' which reproduced the geometry of image formation, wires could be stretched from BX to B and AX to

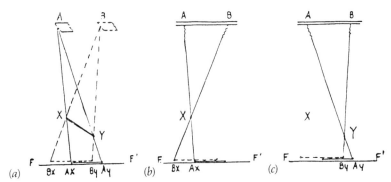

Figure 1.63 The principles of making measurements of distances between anatomical landmarks by the use of stereo films: (*a*) formation of stereo pair; (*b*) location of point X by the cross-wire method; (*c*) location of the point Y by the cross-wire method. The technique originated with Mackenzie Davidson. (From Johnson (1935).)

A, defining X which was then marked in space by a pointer. Similarly (figure 1.63(*c*)) Y was defined. Johnson used the method successfully for assessing pelvic deformity.

This cross-thread method was explained in great detail in a book by Mackenzie Davidson (1916). This book (which was most favourably reviewed by Holland (1916) and came to be regarded as the definitive text on stereo imaging (Mackenzie Davidson, obituary 1919) was of course written during the First World War and the subject matter strongly reflects the influence of the war. The problems described are largely those of locating the geometrical coordinates of embedded foreign bodies, bullets, shrapnel and shot, and the technical problems for the radiologist were in relation to guiding the surgeon optimally. Particularly important was the guiding of optical surgery. Mackenzie Davidson described his invention of a device comprising two trestles across which a stretcher could be laid as patient couch. Beneath the stretcher an X-ray tube was located on guiding rails for making stereo films (figure 1.64). Various equipment for inspecting the films was described, much of it with provision for locating coordinates via the stretched-thread method, such as shown above. Alternatively the surgeon could refresh his memory during operation by viewing pairs of films on a Wheatstone stereoscope which was the common equipment of the day. This comprised simply angled mirrors arranged in a V, and at right angles, each viewing a single film. The observer placed his nose to the V apex such that the right eye viewed the right mirror and vice versa. Such stereo viewing was described (for optical photographs) as early as 1838 by Wheatstone (1838, 1879). Another early pioneer was Brewster (1856). Wheatstone (1802–75) of course never lived to see X-ray stereo imaging.

Figure 1.64 Mackenzie Davidson's stretcher table for the location of bullets by the use of stereo X-ray exposures. The X-ray tube was carried on rails beneath the table. (From Mackenzie Davidson (1916).)

The use of stereo pairs to make measurements by the cross-thread technique, described above, pioneered by Mackenzie Davidson in 1898, does not appear to have been widely used if Manges' account is to be believed. Manges (1911) writes, 'It is difficult to find further literature and it would seem to be a subject of little importance.' He was referring specifically to making measurements of the bony pelvis for obstetric purposes. • Manges describes how stereo pairs can be used to make tracings onto plain white paper of important bony landmarks such as the tuberosities of the ischii, the spinus processes of the ischii, the inner surface of the pelvic brim and the coccyx. This tracing was then used with the cross-thread technique to mark in space the real three-dimensional location of these landmarks between which measurements were then made. Figure 1.65, taken from his paper shows the equipment which Manges used. In discussion of his paper two comments were made: 'I regard the stereoradiogram as the only correct way of measuring, and I am glad that Dr Manges has emphasized that fact in his paper' (from Dr Stover) and 'I think that this method of measurement is not only decidedly new, but very ingenious and exceedingly valuable' (from Dr Johnston). Perhaps it is a comment on the times that such a statement should be made that the technique was 'decidedly new' when it was 13 years old. Manges paper is interesting in another respect for the light it casts on how hazardous this procedure might be. He wrote, 'The question of danger to the foetus may be raised, but we think not fairly. Long exposures are uncalled for, since we have powerful generators,

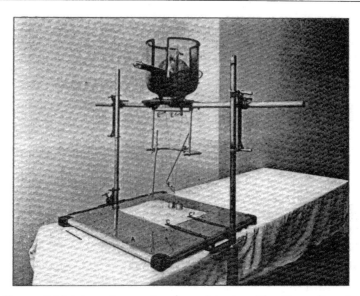

Figure 1.65 Manges' apparatus for pelvimetry. The X-ray films (or double-exposure film) are placed on a table A. The points E and E' mark the positions occupied by the X-ray focus at the two exposures. The threads from these points to the film can just be made out crossing at the point F. At the end of the threads were lead weights. The cross-thread method of distance estimation is described in figure 1.63. (From Manges (1911). Courtesy of the American Röntgen Ray Society.)

extremely sensitive plates and for the very cautious, intensifying screens . . . the process is free from danger.' He also thought all women should have this test before conceiving! 'Furthermore, it would require but few generations to educate prospective mothers to resort to such a measure at some pre-conceptive time.' The technique appears to have impressed Haenisch (1911) on his American journey.

The cross-thread technique for localization and measurement of the distances between internal structures seems to have been reinvented several times. For example the whole procedure is rewritten as if new by Kimble (1935). This paper suggested a large number of applications including measuring the size of abscessed lung cavities and tumours of the lung, width of facial bones, width of the thymus gland, size of tumours, depth of kidney stones, localization of foreign bodies in the eye in addition to the well known pelvimetry.

During the Second World War the problem of locating embedded projectiles became prominent once again. McGrigor (1939) felt that the Mackenzie Davidson cross-thread method was really only of historical interest, and wrote, 'It is not issued for use now. . . . The apparatus, although theoretically accurate, can very easily be mishandled by allowing loose threads and inaccurate measuring by the scales and pointers.' He felt it was useful for teaching purposes but that in the field a variety of methods involving orthogonal radiographs and shadow shift techniques were preferable.

Chausse (1939) developed a stereoradiographic apparatus whereby, being guided by a lever assembly connected to a viewing frame on which was placed a conventional radiograph, a small region of interest in the body could be selected via collimators for the generation of stereo-radiographs with less scatter. He called this technique 'antidiffusion stereoradiography'. The viewing assembly, when used to inspect stereo pairs was further converted into a calibrated 'seriescope' with the same functions as Cottenot's serioscope, except for the improvement due to the diminutive film size. When combined with tomography, Chausse's device performed 'stereostratigraphy'. There is no doubt that Chausse thought that the unpopularity of stereoscopy was due to the poor quality of the pairs (or quartets) of films rather than to some inherent geometry and that his equipment overcame these limitations.

All of the equipment described so far made use of films exposed with shifts applied to the X-ray tube and corresponding lateral shifts applied to the films at the viewing stage. The principles of simulating three-dimensional vision via Röntgen stereoscopy have been described by many workers including Ferguson (1936). This simple paper concludes, 'The satisfaction experienced when we are able to see around and between the various objects, when we are able to differentiate between the anterior, middle and posterior portions of the objects and study them as they stand out in relief and solidity, and when we are able to arrive at a definite decision, more than compensates for the time and care required in stereoscopic work.' The effect of reversing stereo pairs causing a reversal of the orderly manner of all objects in the stereogram (discovered in 1838 by Wheatstone) were also explained and discussed. Meservey (1938), however, provided a deeper understanding of the psychophysics of viewing stereoradiographs.

Jarre and Teschendorf (1933) noted the vast literature on stereo imaging even at that date and the great popularity of the technique in the USA. Their paper, however, set out to clarify the principles which, in their opinion, were not generally understood. They observed the following.

(i) There are important differences between stereo viewing with optically visible and X-ray photons.

(ii) Many observers inherently lack stereo-viewing capability.

(iii) X-ray stereo films may easily be wrongly viewed, creating an apparently satisfying but wrong three-dimensional impression.

Regarding (iii) they particularly noted that it was only when the conditions for the production and viewing of *tautomorphic* images were arranged that a three-dimensional image, accurately representing the object in all spatial parameters (size, angles and location) was obtained. Anything less than this is termed homoiomorphic stereoscopy. The conditions for tautomorphic stereoscopy are as follows.

(*a*) Observation must take place at a distance from the films equal to the distance of the source from the films at exposure.

(*b*) The tube shift must be identically equal to the interpupillary distance of the observer (median value, 65 mm).

(*c*) The central ray from each exposure must strike the film normally.

(*d*) The left-tube shift film must be viewed with the left eye and vice versa.

(*e*) Films must not be inverted about themselves.

It is perhaps because some kind of three-dimensional impression is obtained when some or all of these conditions are ignored, that mistakes can occur. Some common ones are as follows.

(1) If conditions (*a*) and/or (*b*) are violated, this leads to a change of scale in the three-dimensional image, sometimes referred to as a modelling effect.

(2) If condition (*c*) is violated, then no spatial inversions or reversals arise but scale changes lead to distortions.

(3) If condition (*d*) alone is violated and the left-tube shift film is viewed with the right eye and vice versa, a 'pseudostereoscopic' image is formed. The three-dimensional image is laterally distorted and front–back reversed in relation to the true image. According to Mackenzie Davidson (1916) this phenomenon was described by Wheatstone in 1838. Mitchell (1910) possibly noticed the effect radiologically for the first time, writing, 'I made the observation, of which I have never seen any mention in the literature of radiography. . . . The explanation is rather difficult to understand.' Case (1912) made use of the effect to turn an anterior–posterior direction of viewing into the reverse.

(4) If condition (*e*) alone is violated, a 'pseudostereoscopic mirror script' three-dimensional image is formed. This is distorted, left–right inverted and front–back inverted.

(5) If both conditions (*d*) and (*e*) are violated, a 'simple mirror script' three-dimensional image is formed in which front and back are correctly placed but left and right are reversed.

Jarre and Teschendorf cite contemporary equipment able to achieve tautomorphic stereoscopy, for example the Hasselwander Stereoskiagraph and the Beyerlen stereo-orthodiagraph.

Returning for the moment to Mackenzie Davidson, we have already noted that the First World War escalated the interest in radiologically determining the geometrical coordinates of foreign bodies. It is interesting that the publication of Mackenzie Davidson's book almost coincided with the publication of Baese's patent (see § 1.2.12) and we recall that the principal aim of the equipment patented by Baese was the location of projectiles. One might speculate whether Baese's work was known to Mackenzie Davidson at that time. He does not mention it but one must recall that the style of book writing was rather different and that his book only contains a handful of references in total. Locating projectiles was important. Barclay (1949) wrote, 'During the Wars . . . the radiography of fractures and the localization of foreign bodies were not only the major but almost the sole activity of most radiologists.' Some methods which might be considered rather outrageous today were described by Mackenzie Davidson. We might digress for a moment to look at these. They included so-called 'telephone localization' and 'magnetic localization'. In the former the surgeon wore a telephone headset. One wire from the headset was attached to an electrode in contact with a fixed point on the patient. The other was attached to an electrode joined to the surgeon's instrument of incision. When this touched the buried projectile, the surgeon heard a loud click. The patient acted as an electrolytic cell. The method originated with Graham Bell. In the latter technique introduced by Bergonie, a low-frequency alternating current (about 50 Hz) excited an electromagnet placed near the skin, but not in contact. When the field embraced the projectile, it vibrated and the vibrations were detected at the skin surface by palpation. It was claimed that the technique also worked for non-ferrous metals as a result of eddy current induction. The method was used serially interspersed with surgical incisions to guide the direction of cut. Short bursts of power were required to avoid heating the projectile!

The influence of the First World War on radiology was immense. An excellent picture can be formed from Barclay's (1949) account of the first 50 years of radiology. Before the War there was little standardization in X-ray tube design. Gas tubes were unreliable and unsatisfactorily protected and the design of high-tension circuits left much to be desired. The First World War highlighted the importance of X-ray work to the general public and whereas, before the War, X-ray departments were often somewhat back-room affairs, afterwards there began a period of widespread recognition; radiology came out into the daylight as an accepted part of hospital organization. Before the War, anyone possessing an X-ray set could call themselves a radiologist. It was the expectation that a number of men with experience in army life of X-ray work would set up as radiologists that led to a system of formal qualification coming into being in 1921 (Jupe 1946). The regulations and syllabus for this Cambridge Diploma in Medical Radiology and Electrology were published in 1919 (ARE 1919) and a full account of the impact of this qualification may be found in Burrows (1986) and Moody (1970). The importance of bringing physicists and radiologists closer together was a prominent theme of Lord Rutherford's First Mackenzie Davidson Memorial Lecture (Rutherford 1919).

Real-time stereo viewing is also an old technique. Snow (1939) attributes its invention to Caldwell in 1902 and acknowledges the pioneering work of Caldwell, Snook, Kelly, Coolidge, Waite and Rodriguez. The method is described in Mackenzie Davidson's book where he wrote of having set up apparatus, with a double-focus X-ray tube and a 10 Hz shutter between the fluoroscope and the eyes, at Charing Cross Hospital. He observed in 1916 that the method was somewhat problematical but expected it eventually to prove very popular once the technical difficulties were overcome.

The equipment for real-time stereo viewing had to comprise a double X-ray source placed 2.5 in apart in front of a fluoroscopic screen. Either two separate small tubes were used or a double-focus single tube. The current would first activate one tube and then the other, thus forming on the screen a double image. The viewer's left and right eyes would be synchronously permitted to view the alternating images by some blinkering shutter arrangement interposed between the screen and the viewer. The method was also described by Pirie (1911) who used a disc with two holes in it, rotating in synchrony with the Snook apparatus firing two X-ray tubes, and placed between fluorescent screen and observer. Ross (1946) reports the reinvention of this kind of technique.

Snow (1939) developed an interesting modification by which, instead of the shutter, a pair of polarizing filters were rotated between the screen and the observer. The axes of polarization were orthogonal. The viewer wore polarizing spectacles and, by this arrangement, alternate tube images were sequentially presented to the observer with the three-dimensional stereo effect. Snow reported the need to eliminate fluoroscopic lag for successful operation. Very similar principles guided the work of Stamm (1941) whereby large polarizing screens at orthogonal polarizations were placed in front of stereo films. The screen—film combinations were placed at right angles to illuminate a semitransparent mirror at 45° to both. The observer (or more importantly any number of observers) then viewed this mirror with polarizing spectacles obtaining the illusion of three-dimensionality.

Sir James Mackenzie Davidson, pioneer of stereoscopic X-ray imaging died in 1919 (Mackenzie Davidson, obituary 1919). In the last year of his life he had lectured imaginatively to the Royal Society of Medicine (Mackenzie Davidson 1919), demonstrating the principles experimentally rather than with detailed mathematics. His interest in using X-rays may have stemmed from a visit which he made to Würzburg in 1896 to speak with Röntgen (Mackenzie Davidson 1914, Mackenzie Davidson, obituary 1919, Reynolds 1961). He spoke little German and Röntgen little English and Latin proved a common denominator. It was at this meeting that, when asked by Mackenzie Davidson what Röntgen thought of his experiments, he replied, 'I did not think I investigated.'

1.4 Historical perspective by Ziedses des Plantes

The name of Ziedses des Plantes has run like a thread through the course of the early story of body-section imaging. Even if it is disputed that he was the earliest worker in the field, it is quite beyond dispute that his contribution was of major importance. Professor B G Ziedses des Plantes was born on 7 January 1902 in Klundert. He studied electrical engineering at the Institute of Technology, Delft, during 1920–1 and between 1921 and 1928 studied medicine at the University of Utrecht. After specializing in neurology and psychiatry, he was Professor of Radiology at the University of Amsterdam from 1953 to 1972. To celebrate his seventieth birthday in 1972, his former students compiled a volume of his selected works (Ziedses des Plantes 1973). In addition this work listed over 100 of his important papers. It was remarked that all his work was designed and developed by himself and that virtually every thought he expressed on paper was his own. He received many accolades, not least Honorary Membership of the Netherlands Society for Radiology (in 1972). He received the Schleuszner Röntgen Award for planigraphy in 1935.

Earlier in this chapter we have had cause to write often of the remarkable fact that a number of workers were actively pursuing the principles of body-section imaging independently and without knowledge of the existence of each other. This scenario certainly characterized the 1920s although by the early 1930s the connections between the various techniques proposed were being established and the tangled chronology was being worked out. Ziedses des Plantes is still scientifically very active in his eighty-sixth year and has kindly provided a new historical perspective on the early events (Ziedses des Plantes 1988a).

First regarding terminology, the name tomography was chosen by an International Committee during an International Congress of Radiology. The Committee comprised Vallebona (Italy), Ziedses des Plantes (Holland), Stieve (Germany), de Vulpian (France) and Watson (UK). Vallebona preferred the name 'stratigraphy' and Ziedses des Plantes preferred 'planigraphy'. To stay impartial the choice was for 'tomography' and the name was adopted by ICRU in 1962 for *all* forms of body-section radiography. This was of course also the name used for the specific pendulum device by Grossmann. Of contrasting seriousness an interesting anecdote is told by Bricker (1964) who quotes Speed and Brackin (1955) that 'the tomogram was first used by Sir John Tomes, an English dentist, in 1889 and named after him'—a remarkable achievement given that

Röntgen did not make his discovery until 6 years later (Tomes 1897). Perhaps considering his importance for the history of tomography it is worth recalling that the name preferred by Bocage himself was 'radiotomie' (Bocage 1938).

The role of Grossmann in the history of tomography has been clarified by Ziedses des Plantes. In 1932, Ziedses des Plantes showed the results of body-section radiography to Grashey (figure 1.66), then Editor of the *Fortschritte auf dem Gebiete der Röntgenstrahlen*. He introduced Ziedses des Plantes to Holfelder in Frankfurt who was very interested in the radiological technique. When Ziedses des Plantes visited Holfelder, Grossmann was invited also as the director of Sanitas in Berlin. Holfelder requested Grossmann to manufacture a device and Grossmann answered that it would be impossible to make such a device. Some years later, Grossmann manufactured the Tomograph.

Figure 1.66 Rudolf Grashey, Professor at the Burgerhospital, Köln am Rhein. Photograph taken on the occasion of the Second International Congress of Radiology in Stockholm, 23–27 July 1928. (From Renander (1928).)

Ziedses des Plantes has also been able to clear up the confusion noted earlier as to what equipment exactly Bartelink developed. He asserts that Bartelink studied planigraphy from the outset without knowledge of existing patents and the work of Ziedses des Plantes. By coincidence, Bartelink and Ziedses des Plantes presented their results at the same conference session of their Society (see § 1.2.7) in 1931. Remembering the confusion noted in § 1.2.7 it might be fair to say that the association of Bartelink's name with stratigraphy is probably due to Andrews (1936) who stated that Bartelink's technique was based on the stratigraphic principle like Vallebona's. It has already been discussed in § 1.2.7 that all of Bartelink's papers and his patent associate him only with planigraphy.

We may recall that Grossmann had several criticisms of the planigraphic technique; he stated that it had not been used in the clinic (prior to 1935) and that the non-linear motions were unnecessarily complicated. Ziedses des Plantes disputes both with good cause. There is evidence in his thesis (Ziedses des Plantes 1934a) that he used the planigraphic technique extensively for neurological work (prior to 1934) but certainly also acknowledged the spectacular results of Chaoul and Grossmann applying the method to the respiratory system. On the other hand, Ziedses des Plantes drew attention to a recent paper (Ziedses des Plantes 1971) where the clear superiority of pluridirectional motion was demonstrated. Other papers (Ziedses des Plantes 1964, 1978) also drew attention to this and there would appear to have been a resurgence of interest in complex motions at these later dates (Ziedses des Plantes 1973).

Commenting on the work of Kieffer, Ziedses des Plantes considers that his technique using circular trajectories for source and detector was really oblique tomography rather than transverse axial tomography as proposed by Watson. Further comment on the historical perspective is to be found in a letter from Ziedses des Plantes to W H Oldendorf reprinted by Oldendorf (1980).

Turning now to seriescopy, considerable new light is shed on the sequence of events. Ziedses des Plantes (1935, 1938) described the technique in 1935 and again in 1938 and was clearly the originator. Cottenot was not party to the invention. Perhaps it is unfortunate that, as was the style of many papers at the time, Cottenot's (1938) paper did not contain any references. The Messrs Massiot visited Ziedses des Plantes (see § 1.3.2) in 1936 and were shown his seriescopy apparatus. They were requested to construct a new seriescope which they did. They lent a second prototype to Cottenot for application and trial and this is where Cottenot joined the story.

The collected papers (Ziedses des Plantes 1973) corroborate Ziedses des Plantes' claim that he invented planigraphy without knowing of others' work. It would appear that he only came to know of the existing patents when negotiations were well in hand for commercial development and that these negotiations had to be stopped in favour of the device being built in his own workshops in view of the patent protection. It would also appear (Westra 1972) that some manufacturers held the view that the method was of no practical use and tried to dissuade him. Ziedses des Plantes also pioneered several modern techniques which are not always attributed to his innovative work. He invented subtraction

105

angiography and published in 1934 although it was several decades before the technique became popular. The reasons for this he explained in a letter reprinted by Oldendorf (1980). He is also well known for the technique of imaging all ventricles with very small amounts of air, known as the 'somersaults of Ziedses des Plantes'. Perhaps least acknowledged of all is that, a year before Cassen and Mayneord (independently) developed automated radionuclide scanning (Cassen *et al* 1951, Mayneord *et al* 1951, Laughlin 1983), Ziedses des Plantes (1950a) built and used a device for automatic mapping of two-dimensional distributions of radioiodine in the thyroid. He called this method 'indirect autoradiography' (Ziedses des Plantes 1950b). The apparatus is shown in figure 1.67. This link with the quite different topic of radionuclide imaging (see chapter 6) is a convenient place to begin to draw to a close this review of the early beginnings of body-section radiography.

Before this, however, we may note that Ziedses des Plantes has recently provided a brief overview of early developments in section radiography with some wry humour (Ziedses des Plantes 1988b). The paper is the text of a lecture in Udine on 12 September 1987 on the

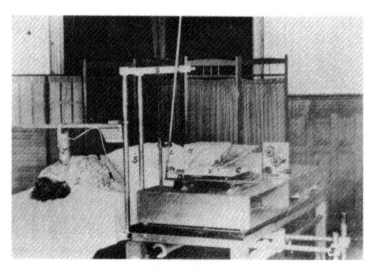

Figure 1.67 The apparatus for 'indirect autoradiography' developed by Ziedses des Plantes and first described at the 1950 (London) International Congress of Radiology. T is a Geiger–Müller tube of length 5 cm and aperture 1.5 cm moving at constant speed in relation to (the thyroid of) a patient. The movement is conveyed to the carriage C by the lever L. A stylus P was attached via a demagnifying arrangement M arranged to create a tracing on moving paper V. An electrochemical recording method was used with a lead stylus and paper moistened with a solution containing potassium iodide, potassium bromide and dextrene. (From Ziedses des Plantes (1950b).)

subject 'Is creative work the privilege of experts?' He uses the history of tomography to show the dangers of believing that it is. It is noted that the originators of the techniques were almost exclusively not X-ray technologists but instead radiologists (Mayer and Vallebona), technicians (Portes and Chausse), a manufacturer (Pohl), a dermatologist (Bocage), an X-ray technician (Kieffer) and a medical student (himself). Moreover his own developments were regarded as impractical by 'experts', such as his own teacher of radiology and not interesting to manufacturers who were more concerned with the existence of certain patents whether they had been applied or not. Ziedses des Plantes wrote, 'After publication of the first results of body-section radiography an outstanding radiotechnical expert wrote a disapproving article (Van der Plaats 1932).' Another expert wrote, 'Planigraphy is an expensive way to obtain bad radiographs.' (See also §3.2 where this has been called the Swedish definition of tomography.) However, Ziedses des Plantes' philosophy concerning experts was not to be misunderstood in that he freely admitted that more modern radiological techniques were firmly the result of the work of experts. He noted the work of Watson, Frank, Takahashi, Oldendorf, Bracewell, Tetel'Baum, Korenblyum, Cormack and Hounsfield in developing axial tomography and X-ray CT. We shall meet all these in later chapters.

He tells an amusing story. He knew nothing about the properties of X-rays and initially thought to construct an X-ray microscope. Consulting textbooks of physics, he soon learnt this was impossible. So after a student rag and a few glasses of beer he invented the idea of motion tomography. From this he drew two conclusions. First, if he had known about the properties of X-rays, he would never have considered motion tomography. Secondly electronic expert systems cannot be stimulated with beer.

Perhaps Bricker (1964) sums it all up best: 'For such an impersonal and mathematical subject it is interesting how large a personal and emotional element has entered the picture. The X-ray focal spot can describe one of several basic paths. Each of these was defended (and often very warmly) by its proponent whilst other methods were criticized. . . . The nomenclature has been chequered with various people suggesting their own terms. A dozen men filed patents. Claims and counter-claims for credit in originating and elaborating this procedure have been prevalent throughout.'

Chapter 2

The Middle Years,
circa 1940–50

2.1 Introduction

In chapter 1 we saw that interest in three-dimensional X-ray viewing is almost as old as the discovery of X-rays and that stereo X-ray viewing was established in the last century. In contrast it was nearly 25 years before the first inventions of equipment for sharp-plane body-section imaging. Perhaps it is rather surprising that, as we have noted, over a period of some 15 years (from 1920 to 1935) a number of inventors separately laid claim, many with patents granted, to the discovery of how to image in this way. Each of these appeared totally ignorant of the work of the others. From about 1934 onwards, papers began to appear, explaining how these inventions related to each other and how the inventors reacted to hearing of the work of others. The early reviews were not entirely consistent but a fairly clear picture emerged. We have seen that to some extent it depends on one's viewpoint who should be honoured with having invented tomography. Bocage wrote the first patent, Ziedses des Plantes built the first equipment, Kieffer, Vallebona and Grossmann were some of the first to make regular clinical examinations by the techniques. The latter half of the 1930s saw the emergence of a number of workers whose work derived from the principles of others. Their papers have a much more modern approach. New developments were placed in context and their merits discussed in comparison with earlier work. Rather more systematic referencing began to be used. The stage was set for progress towards routine implementation.

Against this background, the period between 1940 and 1950 was one of consolidation. It was of course interrupted by a major War. Much work therefore concentrated on the sharp imaging of embedded projectiles. It was less an era of startling new development and more a time of improving the practical implementation. This was also still the time before the impact of the digital computer, and images were still recorded largely by film or by fluorescent screen. It is not really our purpose here to provide a continuous commentary on developments to the present day— and it has already been remarked that tomography is still in use despite the explosion of modern interest in digital computer-based tomographic imaging with X-rays and other modalities. The purpose of reviewing (more briefly) this period is to emphasize the importance of the work pre-1940 and to show how the developments in the decade beginning in 1940 complemented this. Additionally the decade produced some further pioneering work.

2.2 Documentation of clinical work; the War years

Although Vallebona was performing clinical work as early as 1930, it was not really until about 1935 that tomography was performed regularly in the clinic. This was the time that, for example, the Grossmann Tomograph was commercially available from the firm of Sanitas, Berlin. The work in France and Germany has already been mentioned in §§ 1.2.8 and 1.2.6. In the UK the work of Twining may have been derivative in principle but it was pioneering in its simplicity and much of the debate in the late 1930s centred on the relative merits of the home-made and the commercially available equipment. Work on tomography in the UK started relatively late by comparison with the European and US activities.

In the last week of November 1937 a meeting was held at the Royal Society of Medicine to discuss the clinical value of tomography (McDougall 1937, Twining 1937b, BMJ 1937). McDougall was working with the German instrument and Twining with the home-made apparatus for planigraphy. Both were very interested in imaging the thorax although noted that tomography was not limited to this application. McDougall thought that the commercially available equipment was over-complicated and expensive and considered Twining's apparatus sufficient. He concluded his paper, 'I am sure Dr Twining will join me in issuing a warning to the effect that we must still regard tomography as a very young, if somewhat precocious child; but there is little doubt that the boundaries of radiological investigation . . . have been expanded with the advent of this new device.' Twining presented his equipment, noting that it was in commercial production and followed with a series of clinical cases. These were reviewed in the *British Medical Journal* (BMJ 1937) the same week. The review described McDougall as the pioneer of tomography in the UK. He was cited thus by Kerley (1937) and his use of the Grossmann Tomograph was reported as early as mid-1936 in the *Lancet* (McDougall 1936). The review in the *British Medical Journal* commented on the urgent need for simple cheap equipment for routine work. The review also advised caution that 'it would be absurd that tomography could replace the older methods especially stereoscopy'. This view, however, is somewhat contradicted by McDougall (1940) who wrote, 'It has to be recognized that, despite many attempts to stimulate and maintain interest in stereoscopic work, this method of investigation has never been very widely adopted.'

The Grossmann Tomograph and Twining's work stimulated much interest and helped to resolve many of the clinical difficulties of the day associated with pulmonary tuberculosis. McDougall and Crawford (1938) reported clinical cases at Preston Hall using the Grossmann machine and showed how many diagnoses were quite impossible before tomography. Their work there is dated from March 1936. In this paper they refer to similar work done by Colyer (1937) who, like Twining, developed simple apparatus for planigraphy. In Colyer's equipment the tube and detector moved vertically and generated vertical sections of the upright patient. The clinical importance was that lines of fluid levels in cavities could be established.

Colyer (1937) described how he came to be interested in tomography. He had a private patient who was associated with the management of Preston Hall where McDougall had installed a Grossmann Tomograph. Colyer had read of the work of Bocage, Grossmann and Twining and was also of the opinion that the Grossmann Tomograph was too costly. Perhaps inspired by the 'home-made' approach of Twining he made a similar development but enabling tomography in the vertical position. He wrote, 'I made another apparatus which I have been using now for some months and with which I have obtained the results I expected. It consists of an ordinary upright sinus stand fitted with a counterbalanced Potter–Bucky, in conjunction with a shock-proof rotating-anode tube, mounted on a firm stand which moves along a track fixed to the floor. The cross-arm of the tube stand is coupled to the Bucky by a wooden lever, which rotates about a pivot fixed to a bracket projecting from the wall. The position of the pivot is variable and determines the depth of the layer examined. The action of the rocking lever is to move the tube and Bucky in geometrical ratio in opposite directions. . . .' Colyer installed this apparatus in the Purley and District War Memorial Hospital.

A new method of stratigraphy was introduced by Ronneaux *et al* (1939). The equipment, called the Oscillostrator, was manufactured commercially by the Companie Générale de Radiologie. It is shown schematically in figure 2.1. A fluorescent screen (on the left) views the beam from the source (on the right). In turn the screen itself is viewed by a camera. The screen and camera combination can rotate about the vertical axis. The idea was that the stationary patient was viewed on the rotating screen until the section of interest was located, whereupon

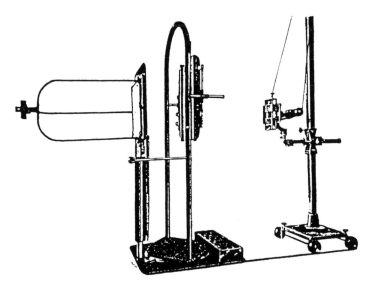

Figure 2.1 The Oscillostrator of Ronneaux, an apparatus for stratigraphy with recording on miniature film. (From Ronneaux *et al* (1939).)

minified images were recorded by the camera. When these were enlarged, their quality was shown to be equivalent to directly recorded large films. The camera was on a mechanism similar to the trombone so that its distance from the screen was variable (generally 0.8–0.9 m). The source-to-screen distance was 1.2 m. The economy of film usage was considered important and the method was attributed to de Abreu. The equipment was used for pulmonary stratigraphy.

The clinical importance of tomography was such that two books which were devoted solely to the subject appeared. McDougall (1940) compiled an atlas of clinical tomograms with only the briefest of introductions. The history of tomography received only a few paragraphs; the principal concern was to gather together an album of photographs, many from Twining at Manchester Royal Infirmary. Indeed the historical information is not entirely consistent with early papers; he wrote, for example, as if Bocage had had trouble with his apparatus whereas there is no evidence he ever produced such. Naud (1936) in his doctoral thesis states that no equipment was built. Possibly McDougall was referring to 'the Biotome of Dr Bocage' built by Massiot (see § 1.2.8). The appearance of this book, however, is important, indicating the status of tomography at the time and is somewhat analogous to the appearance of atlases of CT images not long after the commercial development of X-ray CT in the early 1970s.

In 1946 a much more extensive clinical atlas of tomography appeared

(Weinbren 1946), showing the maturity of the subject during the War years. Again historical aspects received very little attention although it was noted that in 1933 and 1935 the editors of the *Yearbook of Radiology* were unenthusiastic about tomography—'an interesting technique of doubtful practical value'. In 1936 they wrote, 'This ingenious method . . . appears to have some value . . . where other methods fail.' The introduction to this book noted that one of the first, if not the first, complete tomograph in the UK was installed at Queen Mary's Hospital, Roehampton, where it was used by Dr Helen Harper. Dr Harper went to Berlin in the 1930s to see the Grossmann Tomograph and persuaded the Ministry of Pensions who then administered Roehampton Hospital to buy the apparatus. The radiographer at the time was Watson who, as we have seen, became well known for his own development of transaxial tomography (Prior 1988). After Weinbren's atlas it was not until 1969 and 1973 that other really extensive clinical textbooks on tomography appeared (Takahashi 1969, Berrett *et al* 1973).

In the USA, Kieffer (1943) continued to expand on the use of laminography. This was a detailed paper like an earlier one (Kieffer 1939b) in which he had analysed the factors controlling the quality of the images in laminography. He once more pointed out the independence of his work from earlier published studies. His writing, like that of Grossmann in the 1930s, was scientifically detailed; he considered the advantages of different types of motion and also nicely showed how planigraphy and vertigraphy (detector perpendicular to the plane of motion of detector and source) were in essence special cases of laminography. This becomes clear from inspecting figure 2.2. He credits Andrews with the first experimental activity in the USA.

We have already seen that stereoscopic imaging was much employed before body-section imaging became widely known. A glance at the contents of Muir's (1924) practical manual of X-ray work shows a heavy concentration on stereo work, largely for the location of foreign bodies. 30 years later, however, further volumes on stereo imaging were still appearing, such as that by Hasselwander (1954). Hopf (1943) introduced a new apparatus for stereoröntgenoscopy. Wiegelmann in Germany introduced stereoscopic screening in 1946 (Ross 1946). Two X-ray tubes were energized alternately at 50 Hz, casting two images on a fluoroscopic screen. The viewer wore spectacles with each lens equipped with an alternating shutter operating in synchrony with the X-ray tube. The equipment was much in use for guiding surgical operations such as fracture setting. The use of shadow-shift techniques for localizing the depth of projectiles was once again topical in the Second World War years as it was in the First World War. Scott (1940) showed how a simple portable device could assist the process. The device was a strip of metal with holes drilled at regular spacings. The strip was placed at 45° to the

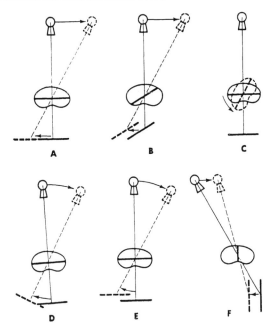

Figure 2.2 The techniques for body-section imaging: (*a*) planigraphy; (*b*) laminography; (*c*), (*d*) stratigraphy; (*e*) tomography; (*f*) vertigraphy. (From Kieffer (1943).)

object being radiographed and a shadow-shift film was exposed with the tube motion (of arbitrary travel) in the direction normal to the strip of metal. Whichever hole gave the same linear shadow shift as the embedded projectile was clearly at the same depth, which then became known. At the end of the War the opportunity for the Allies to study captured German equipment was taken (Allen *et al* 1966) including a device being tested in the University of Erlangen invented by Professor Albert Hasselwander for the stereoscopic localization of foreign bodies. A semitransparent mirror replaced the usual completely reflecting mirror in the stereoscope and a pantographic pencil was used to sketch the outline of the projectile on a cross-section of the part by the viewer employing stereoscopic vision (figure 2.3).

Further evidence of the clinical importance of body-section imaging was provided by Baylin (1944). He showed that its value lay not only in the then well established areas of chest imaging and the study of skeletal joints but also in the paranasal sinuses, soft-tissue tumours along the spine and kidney tumours. The technique was of inestimable value to the thoracic surgeon for detecting, for example, cavities in the post-operative chest and delineation of the bronchial tree. Baylin said that, for imaging the temperomandibular joint, laminography totally replaced conventional

113

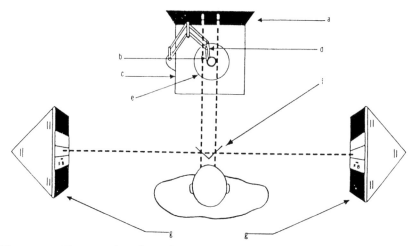

Figure 2.3 The Hasselwander technique for the stereoscopic localization of foreign bodies by means of cross-sectional drawings. (*a*) Clack screen; (*b*) pencil; (*c*) paper; (*d*) light marker; (*e*) drawing of cross-section with foreign body; (*f*) semitransparent mirrors; (*g*) illuminators. (From Allen *et al* (1966).)

planar imaging. His work was one of many cited in the 1940s in the *Yearbook of Radiology*.

Didiee (1944) provides a fascinating view of the status of section imaging in France during the War. After reviewing the history of section imaging, he says that the number of pieces of equipment in France at the time of writing could be counted on the fingers. The War had led to a dramatic interruption of scientific progress as well as to shortage of supplies of consumables such as X-ray film. Despite this, there had been some progress in bringing to bear section-imaging techniques in the clinic particularly in pulmonary imaging. Didiee argued that, although the theoretical basis of section imaging was well understood, it was still possible to obtain a confusing diagnosis from the section images. He advocated the use of section imaging from several directions of cut before conclusions were drawn. He suggested two means of making progress with the subject: firstly that what literature there was should be reviewed *in toto*—and he regarded the literature of variable quality—and secondly that more experimental work should be done with phantoms. Didiee's own paper addressed this latter point. He created both direct and section images of a phantom comprising a wedge of gruyère cheese which he considered to have cavities similar to pathological lung tissue. His results certainly demonstrated that the sectional images were an improvement on the direct images but unfortunately the sectional images also showed false structure that did not correspond to the cut-up cheese. Hence he suggested great caution with the technique. The movements he employed were unidirectional, multidirectional and epicycloidal.

A year later McDougall (1945) was once again lecturing on tomography. He noted that, in recent years, stereoscopy had received revived interest but thought that even in expert hands the method was limited. Also, of course, many people lack stereoscopic vision. McDougall presented case studies of imaging the thorax, the site in which he had most experience. In general he recommended the use of 2 cm spacing between sections. He wrote more conservatively of other body sites, '. . . experience . . . is scarcely wide enough to warrant a description of results and to enable us to come to any decision upon its value in such conditions.' Moreover, 'I have no desire to make any exaggerated claims for tomography, and the indiscriminate use of the apparatus in all cases is to be deprecated.' However, he commented that it may well find application in the cranium, in cases of Hodgkin's disease and in tomography of the heart. Although this paper makes no significant historical comment, he draws a comparison between the apparatus of Grossmann and that of Bocage, noting that in the former the slice width was finite, dependent on the arc of travel, whereas in the latter the slice was infinitesimally thin. Presumably he referred to embodiments of Bocage's principle (such as by Ziedses des Plantes) since, as far as can be told, Bocage did not construct apparatus.

Tomography appears to have become established in Denmark in the late 1930s; A Oberhausen, a graduate engineer and managing director of Sanitas, lectured on tomography to the Danish Radiological Society in 1938 and F Moller to the Medical Society in November 1942 (Moller 1968).

The influence of the War on radiological practices has been the subject of comment already. An intriguing article appeared in 1945 showing how a kit of parts could be prepared for carrying from station to station to be attached to conventional X-ray equipment to perform planigraphy (Brock 1945). The equipment was simple, inexpensive and portable. Brock wrote, 'The parts when separate can be tied in a bundle small enough to fit in a small locker and be carried as baggage. During the present emergency many radiologists in the Armed Services would like to have the additional information of body-section röntgenography but do not feel that they can construct a new planigraph on each change of station or for each new installation of equipment. This apparatus has accompanied the author on all of his changes of station with no more difficulty in transportation than is found with moving textbooks.' The equipment built into a horizontally disposed series of levers, rather like Twining's apparatus, was found to be suitable for installation on equipment manufactured by Picker, Westinghouse and Keleket. The equipment was engineered at the Department of Radiology at Columbia Hospital, Milwaukee.

War conditions of course posed many problems for field radiology as instanced above. A similar plight was noted in Africa by Goldman (1943) that tomography which was routine in chest hospitals made use of equipment which was quite unsuitable for a military general hospital. It was also felt that tomography in the vertical position was a better way of seeing pleural effusions, cavitations and fluid levels and that conventional equipment did not really allow this. Goldman reported the construction of yet another home-made device of levers for achieving the movements required of the X-ray tube and the cassette (see also §1.2.9) and reported results as good as those of Twining and Bush. Goldman's paper is an interesting insight into the problems of the time. He concluded, 'Our means of studying radiological literature are necessarily very limited, but in the journals at my disposal I have not been able to find a description of a similar appliance.'

Yet another on the same tack was Robin (1945), a major in the US Army. He wrote, 'The X-ray Departments of Zone of Interior Army Hospitals have been supplied with adequate standard equipment. From time to time the need arises for special types of X-ray procedures which can be accomplished frequently by calling upon the Hospital Utilities or the Post Engineers for help in constructing appropriate devices. The apparatus (for body-section radiography) was quickly fabricated by the carpenter shop of the Post Engineers, Station Hospital, Camp Howze, Texas, after a very brief description and working diagram of the requirements were furnished by the X-ray Department.'

Many applications of radiology in the Second World War were rather different from those in the First World War (see §1.3) when the principal attention had been focused on the radiology of fractures and on the location of foreign bodies. Between the Wars there had been dramatic developments in the equipment for producing and detecting X-rays (Barclay 1949). Gas tubes had been superseded by hot-cathode tubes. Films had replaced glass plates and intensifying screens were routinely used. As a result, exposure times were measured in fractions of seconds rather than minutes. There was valve rectification, shock-proof tubes and cables, rotating-anode tubes and non-flammable film (see also appendix 1). When War came in 1939, the position was vastly different from that of 25 years before with adequate radiological planning at least on paper.

A fascinating insight into the practice of radiology during the Second World War comes from the monumental work published by the Office of the Surgeon General, Department of the US Army (Allen *et al* 1966). This report separately considers the individual theatres of operations. Whatever materials were to hand were cobbled together to make inexpensive section-imaging devices referred to generally in this report as laminographs. For example in the Zone of the Interior at McCaw General Hospital, Walla Walla, Washington, Major Gerhard Danelius MC constructed an inexpensive laminograph which 'produced radiographs comparable in quality with those produced by the expensive commercial equipment'. In the Mediterranean theatre the history of the Sixth General Hospital mentions improvising a laminograph by the Ninety-eighth Ordnance Company and fitting it to a 200 mA X-ray set (Langley 1944). In the European theatre of operations, laminographs were also improvised. The method of disseminating information on how to do this was by chain letters which were passed from hospital to hospital. Wherever practical, these were accompanied by models or by photographs and designs. This started in 1942 but by 1944 there were too many hospital units in the theatre and other methods of passing information were used such as theatre medical bulletins and conferences. Ordnance units gave assistance in improvising from scrap materials. As well as improvised laminographs, home-made stereoscopes were also fashioned in most theatres of the War. The use of X-rays during hostilities has received much attention as a study in its own right (see, for example, Reynolds 1945). Surprisingly Hodges (1945) reviewing the first 50 years of the development of diagnostic X-ray apparatus completely fails to mention tomography in any form.

2.3 Gabriel Frank's pioneering patent

Whilst 1940 saw widespread clinical use of body-section imaging by the methods reviewed, it was also the year of a pioneering patent by Gabriel Frank in Budapest (Frank 1940). It would be no exaggeration to say that this patent described a practical method of performing what we today call X-ray CT, some three decades before the time which many regard as the birth of CT. Computational techniques were not sufficient in 1940 to achieve CT and Frank's equipment used an analogue method of reconstruction.

The pioneering aspect of Frank's patent was the recognition that, whatever the method of body-section imaging utilized, the plane which was said to be in focus was not strictly isolated from shadows of overlying and underlying structure. This could never be the case in view of the fact that the X-rays passed through significant structure in other planes, the attempt having been confined to blurring these structures over the unblurred shadow of the plane in focus. Without naming names, Frank grouped all such methods together and described, in the opening of his patent, this major disadvantage.

He described a quite new concept by which the influence of overlying and underlying structure was totally avoided. Today these principles are very well understood by medical physicists. However, the fact that the principles were given as early as 1940 and clearly with a medical diagnostic application in mind makes the case for reviewing what he proposed in some detail.

The figures from this historic patent are shown in figure 2.4. In

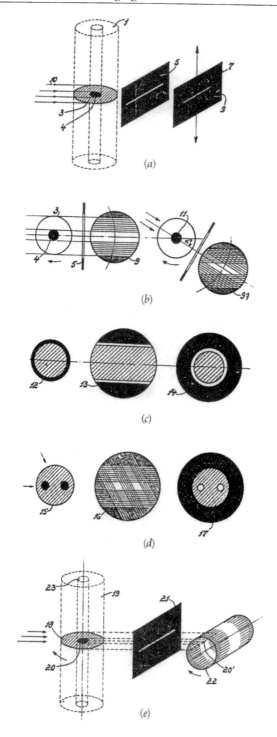

(a)

(b)

(c)

(d)

(e)

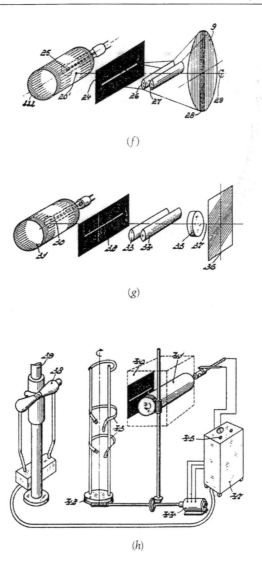

(f)

(g)

(h)

Figure 2.4 The historic figures from the patent by Gabriel Frank showing how a transverse axial section may be sharply imaged without contributions from adjacent sections—in essence the principle of CT without the C(omputer). For detailed description see text. (a) How a sinogram can be formed on a flat plate. (b) How by back-projecting sinograms from different orientations a sectional image is constructed. (c), (d) Examples of the back-projection for two phantoms. (e) The formation of a sinogram on a rolled film. (f) Optical back-projection from the sinogram to yield a cross-section via a rotating-detector stage. (g) Optical back-projection from the sinogram to yield a cross-section without rotating the final detector stage. (h) Frank's experimental arrangement for generating the sinogram. (From Frank (1940).)

119

figure 2.4(a) is shown a section 3 of an object 1 which it was required to image. X-rays 10 in a parallel beam from a broad focus were directed at the section and the exit beam intercepted an absorbing screen 5. The screen contained a slit which passed a narrow bundle of the rays onto a detector 7. If this detector were linearly translated as the object rotated, a series of views of the object would be laid down upon the detector. In figure 2.4(b) are shown these projections of the slice for two orientations of the source–detector assembly relative to the stationary body. The implication is that, if these line shadows could be back-projected onto a plane, an image of the section would result after many orientations—the familiar CT principle. Figure 2.4(c) and figure 2.4(d) illustrate this for a head and an arm phantom respectively. Frank was quite clear that the image which was formed was not suitable for direct perception in the way that the previous tomographic methods were able to provide. The remaining drawings concern how to reconstruct from the recorded projections. In figure 2.4(e) a light-sensitive film 22 is wrapped on a cylindrical former such that the slit in the screen 21 is aligned with the axis of the cylinder. The X-ray source is now stationary and the data are taken by simultaneously rotating the object containing the section of interest and the cylinder around which the film is wrapped. The film, after a complete rotation would contain what we today call a sinogram of the slice to be reconstructed. Reconstruction proceded via an inverse procedure (figure 2.4(f)). The film was wrapped around a cylinder 20 at the centre of which was a source of light 25. The source illuminated the whole film uniformly but only the shadow of that part of the film, a line, in the direction of a screen 24 was passed through the slit in this second screen. The narrow bundle of light rays thus defined were then expanded onto another sensitive plate or film 28 via cylindrical lenses. This was to achieve optical back-projection from a line onto a plane. (For a discussion on a possible error in Frank's patent see the book by Mackay (1984); the cylindrical lenses are wrongly placed to smear out the projection in the direction of the shading.) All that now remained was to rotate the cylinder and the plane of the second detector simultaneously about their respective axes to achieve an image of the section, uncontaminated by other planes parallel to it.

The narrow fan of light rays which emerge from the slit in the screen in figure 2.4(f) and which are incident on the cylindrical lenses are the optical analogue of the X-ray narrow fan which emerges from the slit in the screen in figure 2.4(e). It is required that the dimensions of the roller or cylinder in figures 2.4(e) and 2.4(f) exactly correspond. Then, as the receiving screen or film 28 is rotated in synchrony with the roller 20, the expanded rays emerging from the cylindrical lenses blend together to produce the optical analogue of that slice which was investigated by the X-rays in figure 2.4(e). Frank's apparatus was based on the principle of continuous rotation. The technique rests on all angular rotations of object (or correspondingly source and detector about object), X-ray detector, light-projecting cylinder and final light-sensitive plate or film all rotating at the same angular velocity.

Figure 2.4(g) shows a similar arrangement for reconstruction by optical methods. In this arrangement, instead of rotating the light-sensitive detector, the expanded rays from the cylindrical lenses are rotated by a second lens system 35. The detector remains stationary. If the screen could store the image the observer would gain an impression of the section gradually building up if he trained his eye on the screen during the rotation.

Figure 2.4(h) shows Frank's practical implementation of the invention. The patient is immobilized in the cradle 43 which rests on a turntable 42. The X-ray source 38 is stationary supported by a frame 39. The film 41 is contained in a box which on the source side contains the necessary slit. A simple arrangement of gears enables synchronous rotation of the detector and the patient via a motor 44.

In Frank's patent the beam of X-rays forming the image is parallel. This is difficult to achieve in practice and in general in the modern adaptations of the method of optical analogue reconstruction the beam is a diverging fan. The reconstruction is then similar except that the final recording film must be tilted relative to the beam axis. The theoretical justification for this is clear from, for example, Barrett and Swindell (1977) and one imagines that Frank had not realized this possibility.

The history of X-ray CT is another vast subject. The year 1972 is generally remembered as the time when a commercial medical machine became available. The mathematical principles of reconstruction date to Radon (1917) whilst Oldendorf (1961) made some pioneering experiments in 1961 and records show that a medical X-ray CT scanner was designed in Kiev in 1957 (Koryenblum *et al* 1958, Tetel'Baum 1957, Barrett *et al* 1983). (We shall return to this topic in part 2.) However, in Frank's patent we appear to have a clear statement that, even before the impact of the digital computer, tomography was possible by the method of reconstruction from projections. Since then, and much later, his optical method has been improved upon by many workers (see, for example, Barrett and Swindell 1977, Gmitro *et al* 1980, Lindegaard-Anderson and Thueson 1978, Peters 1974, Edholm *et al* 1978).

There is a serious flaw in the method proposed by Frank which was put right by these later workers. The image reconstructed by Frank's technique is a direct unfiltered back-projection and it is well known that this is the convolution of the true image with a $1/r$ function. (Readers unfamiliar with the theory of CT may wish to consult appendix 4 at this stage where an account has been given in words and pictures.) In order to take account of this need to filter the image from back-projection (or equivalently to filter the projected sinogram *before* the operation of back-projection), it is necessary to resort to the use of bipolar filters. Barrett and Swindell (1977) show how this may be achieved. In their implementation the back-projecting function of the cylindrical lens is also dispensed with. Instead a point beam emerges from the cylinder into a beam splitter (for bipolar filtering) and the rotating cylinder is rocked through 90° relative to the normal to its surface in order to reconstruct all points in space.

The method of Lindegaard-Anderson and Thueson (1978) ignores the filtering function and dispenses with the cylindrical lens. Instead the film detector is translated (for all orientations of the object) to achieve the back-projection. Peters' (1974) system uses the cylindrical lens and an optical post-processing stage to perform the deconvolution of the $1/r$ function.

The apparatus of Edholm *et al* (1978) also bears a striking resemblance to that of Frank. In this the film is exposed flat and lines are picked off one by one by a scanning slit, expanded by a cylindrical lens and imaged onto a rotating film detector. The image generated, however, has been correctly filtered by a modulation applied to the light intensity of the illuminating bulb. Others may be credited with improved implementations but Frank first thought it all out as early as 1940.

2.4 Transverse axial imaging in the 1940s; Watson and Takahashi

In § 1.2.11 we saw how Watson was the first to patent equipment for transverse axial imaging although Kieffer's apparatus was also capable of this mode of imaging. In the 1940s, several other workers were developing similar equipment. Frain and Lacroix (1947a) in a very short paper presented results from their own development. Interestingly the physics and engineering of the development are not mentioned. There is simply one small figure showing how the patient and the film rotate in synchrony in front of a stationary X-ray tube. The principle that one particular plane was imaged in focus was clearly understood and used to image the heart and to measure its diameters. Frain and Lacroix stated clearly that the only other work of this type of which they were aware was that of Kieffer in 1929 and they comment briefly on the differences between their techniques. The paper makes no mention of the work of Watson nor of Vallebona (see §3.1). The work was referred to in two other papers in the same year (Frain and Lacroix 1947b,c). However, a short while later Frain and Lacroix (1948) further exemplified their technique for horizontal transverse axial tomography and claimed that, as they had introduced the first experimental results on 14 January 1947, they were earlier than Vallebona's presentation on 3 March 1947. Once again they referred to the work of Kieffer in 1929 and added Amisano and Herdner. In 1948 they were claiming to be able to generate non-interfering cross-sections with 1 mm separation via the use of grazing

X-rays, 360° rotation and synchronous same-sense co-rotation. Apparatus based on the principles of Frain and Lacroix was manufactured by Massiot (the Radiotome) and exhibited at the First Congress of Radiologists of the Latin Culture in Brussels in 1951 (Massiot 1974). A Massiot apparatus for axial tomography was reportedly in use by Antoine (1952). It used a grazing angle of 20° for the X-rays. The film and patient rotated in front of a stationary X-ray source. Antoine considered the pluridirectional movement to be a distinct advantage over stratigraphic methods. His paper also mentioned Frain and Lacroix, and Vallebona. Equipment based on the ideas of Vallebona had been manufactured by Zuder a little earlier than that of Massiot.

Other workers in this period were (cited by Stephenson (1950b)) de Abreu (1944) and Herdner (1948a,b), Janker (1950), Giacobini and Manzi (1954) and of course Vallebona (1948). Janker (1950) published the construction of an apparatus with which it was possible to shoot transverse body sections on the recumbent patient with the tube and film revolving around the patient on a circular arc of 180–250°. Janker only demonstrated the principle (Gebauer 1956) and the apparatus was realized by Giacobini and Manzi (1954).

Vallebona (1948) cited the work of Frain and Lacroix, de Abreu, Kieffer and Amisano. In this paper he presented a diagram of the principle of transverse axial tomography and clinical results of using the equipment developed in 1947 and commercially produced by the Zuder Company of Genoa under the name 'Pantixstrator' (Gebauer 1956). In chapter 3, further light is shed on Vallebona's work from his later publications (see, for example, Vallebona 1950b).

In Germany, Gebauer (1949, 1956) built equipment for transaxial tomography based on identical principles with those patented by Watson; the patient and film co-rotated whilst the X-ray source, inclined at an angle to the horizontal, remained stationary. These papers showed many images from the thorax as well as work with phantoms. de Vulpian (1951) experienced difficulties with radiographing patients in the upright position and proposed new apparatus to overcome this difficulty.

Transverse-section tomography was implemented commercially by several companies including the 'Pantixstrator' (Zuder, Genoa), the 'Transversal Planigraph' (Siemens–Reineger, Erlangen), the 'Transversotom' (Elektrix, Linz) and the 'Radiotome' (Massiot et Cie, Courbevoie) (Gebauer 1956).

Watson (see § 1.2.11) first conceived the idea for transverse axial tomography in 1936. The prototype equipment which he called the Sectograph was constructed just before the War. The base (with the rotating turntables) was made by the Medical Supply Association and 'a grateful patient supplied the superstructure for the patient's support' (Watson 1962). Construction of the machine was encouraged at Queen

Mary's Hospital, Roehampton, by Dr H M Harper who had installed previously a Grossmann Tomograph. The first commercially built apparatus by the Medical Supply Association, also called the Sectograph, was made in 1938. Mr H W Grover, Manager of the Association, not only helped to design and build the first commercial apparatus but also assisted in transporting it to various parts of the country. The Sectograph was lent to hospitals in Birmingham, London and Bristol but Watson (1962) records that 'no useful work appears to have been done with it'. It remained in Bristol for the duration of the War. In 1948, Mr Quick asked if he could have the machine and it was transferred to Weston-super-Mare. Watson (1962) notes, however, that he was on the verge of mastering the technique when an urgent demand for the Sectograph came from the Royal Cancer Hospital in London and in a short time the machine was once more installed there. Work was performed by Dr J J Stephenson and published in 1950 (see §3.1). This and other activity in Europe appears to have led to a blossoming of interest in the technique.

Watson (1943) concluded his series of three papers on differential radiography, read to the Scottish Radiographic Society. (These papers have subsesquently come to be regarded as the internationally recognized thesis on the subject and were important enough to have been singled out for comment in the History of the Society of Radiographers (Moody 1970).) In this paper there are figures of his commercially produced equipment and clinical results taken with it. This paper is additionally interesting for the light cast on practical section radiography during the War. He wrote, 'Nothing has created so much interest as the Grossmann Tomograph . . . and it is customary to call any sectional radiograph, a "tomograph".' He advocated the use of this word for section images obtained by unidirectional curvilinear motion only. The advice appears to have been ignored since the word came into common use for all forms of sectional imaging. Also, 'At the present time, choice of apparatus is very limited.' He said that further developments for his own Sectograph had been delayed on account of the War but that there were 'endless possibilities'. The paper contains two memorable sentences. The first is: 'Used discriminately there is no reason to believe it (that is differential radiography) will ever be discarded or replaced.' We now know that digital CT has replaced analogue section imaging for most applications but from one angle his remark might have been meant to apply to all section imaging, however achieved. The second memorable statement is a prophetic view of the role of section imaging in radiotherapy treatment planning. He wrote (of horizontal sections from the Sectograph), 'Undoubtedly they offer a valuable field of application in the plotting of individual therapy dosage charts.' This role only became really well established after the birth of X-ray CT although pioneering work was done in Japan by Takahashi (see §3.1).

Watson was a key figure in the development of X-ray imaging equipment in the UK. His precise role with regard to the installation of a Grossmann Tomograph at Roehampton (see § 2.2) is not known but it is reasonable to assume that he worked with it and that this experience influenced his own inventions (Prior 1988). He worked for the Ministry of Pensions at Roehampton from the end of the First World War to 1935 when he joined the newly formed Ilford Department of Radiography and Medical Photography headed by Miss K C Clarke at Tavistock House in the British Medical Association building. He remained at Tavistock House until retirement in 1960 apart from a 2 year period 1939–41 when he returned to Roehampton to work for the War Office and the British Red Cross Society. With K C Clarke he was instrumental in producing the 'bible' of radiography *Positioning in Radiography* (Clarke 1939). Mercer (1988) has noted that not only was this book—which includes many details of body-section imaging—produced at a financial loss to Ilford but that indeed it was policy at Ilford that all new techniques and inventions were shared and no commercial gain was made by any individual. Watson had an excellent relationship with the design engineers in the British X-ray industry, which was small at the time, and in particular with the Medical Supply Association who marketed his inventions. Watson had his own workshop at home where he made preliminary experiments in many branches of radiography including body-section imaging. He was the prime mover in axial transverse tomography whose ideas were adapted widely by others. Even after retirement he remained active in developing transverse axial tomography.

2.4.1 Takahashi's experiments in Japan

In 1945 the Japanese radiologist Shinji Takahashi began a remarkable series of developments leading among other achievements to the development of transverse axial tomography. By 1953 he had completed a large number of experimental inventions. He produced 50 or so papers mostly in Japanese and today the most convenient way of accessing his work in English is via his book *Rotation Radiography* (Takahashi 1957). He wrote this mainly because after teaching on the Third International Lecture Course on Stratigraphy in Genoa, organized by Vallebona in 1955, he became aware that the developments from Japan were little known in the West. Although the book did not appear until 1957, many of the experiments it reported were from 10 years previous and firmly belong to the late 1940s. Looking at Takahashi's work today, it is especially remarkable how similar were his final pieces of apparatus to those developed independently by Watson and by Vallebona.

The lack of good scientific interchange between East and West has also been commented on by Goldberg and Kimmelman (1988) in relation to the early history of diagnostic ultrasonic imaging. They wrote, 'The early post-War Japanese work would remain obscure in the West until the early 1950s when international conferences on biophysical and medical ultrasonics introduced Eastern and Western investigators to each others' efforts.'

It is unnecessary to say that Takahashi's activities were prompted by the shortcomings of projection radiography. He knew about tomography of the Grossmann type but was more interested in transaxial section imaging and clearly did not know this had already been accomplished when he started work. He developed unified equipment to execute what he called rotation radiography. The equipment itself was called the Universal Rotatograph and is shown in figures 2.5 and 2.6. With this he was able

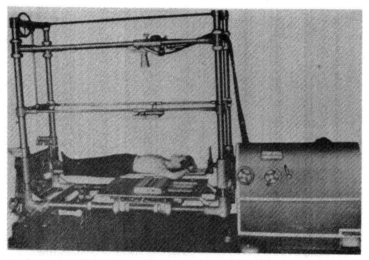

Figure 2.5 Takahashi's Universal Rotatograph. (From Takahashi (1957).)

to perform four separate types of radiography which he called rotation sighting radiography, cross-section radiography, solidography and rotation kymography. In the context of this book it is with the second and third of these that we are most concerned. Takahashi's book is a logical and chronological account of how one development led to the next and at each stopping point he conjures up terminology of his own to describe what was achieved. At first sight some of this terminology is a little confusing and in this account an attempt will be made to relate this to terms which are in more modern use.

Figure 2.7 shows a summary of Takahashi's developments linking them together in this way and indicating the date at which they were made.

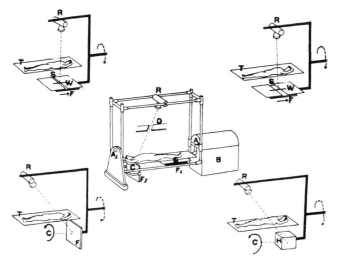

Figure 2.6 Schematic diagrams of the Takahashi Universal Rotatograph operating in various modes for rotation radiography. R is the X-ray tube, D is a diaphragm, T is the radiographic table, S is a lead slit, F is a film, W is a lead diaphragm, C is the rotation centre of the film and H is a box cassette. (From Takahashi (1957).)

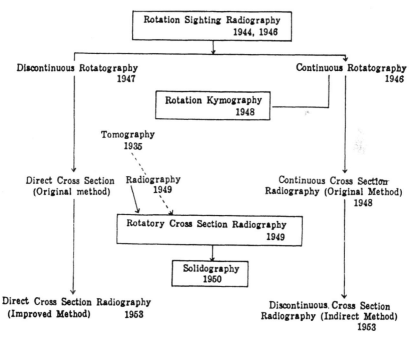

Figure 2.7 A summary of the various developments made by Takahashi with the names which he coined. The dates at which the inventions were made are shown and the techniques are described in turn in the text. (From Takahashi (1957).)

In essence the Universal Rotatograph was an arrangement of supports by which a source and detector could be rotated about a patient who remained stationary. At the same time as the detector rotated about the principal axis, it could where necessary be translated or rotated about its own axis. The need for this will be apparent from what follows. In passing it might be noted that the equipment also allowed Grossmann-type tomography to be performed as well as simple planar static radiography.

Takahashi described the results of several experiments by which the X-ray tube and detector support remained stationary whilst a phantom was rotated between them. The X-rays were collimated to a narrow fan via slits and at each orientation the film was advanced in its holder to expose a new strip. Takahashi's diagram is shown in figure 2.8. He called the resulting film a rotation radiogram; today we would recognize this as the sinogram of the section being imaged. He illustrates the sinograms for a number of objects of increasing complexity, for example a single point, two points, n points, single and multiple straight lines, curved lines, planes and solid objects. As an example of his work, figure 2.9 shows the sinogram for two closed convex curved lines in which it may be noted that one zone on the sinogram remains contained by the other.

We might pass over the first of his applications of the Universal Rotatograph, namely rotation sighting radiography, in that this is essentially a method of making accurate measurements from a series of projections whose relative orientations are precisely known. In itself this was not a form of section imaging. When the beam was confined to a narrow

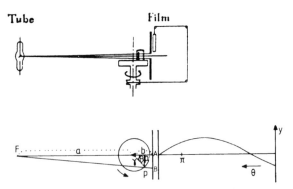

Figure 2.8 A sketch of the apparatus used by Takahashi for model experiments to study the laws governing rotation radiography. The X-rays were collimated to a narrow fan beam via a slit. The film translated behind the slit as the table on which the sample was placed was rotated. This is shown controlled by a pulley wrapped around the table and connected to the film carriage. Today we would call the image which was built up on the film a sinogram. The sinogram of a point absorber is shown to the right of the lower drawing. (From Takahashi (1957).)

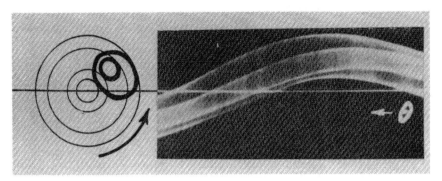

Figure 2.9 The rotation radiogram (or sinogram) of two closed convex curved lines. (From Takahashi (1957).)

fan, distances in the section irradiated were deduced from measurements on the sinogram.

Of much more interest and relevance to this account is Takahashi's second use for the Universal Rotatograph: rotation radiography. Work began in 1946 and he classified the stages of development under seven headings. Without wanting to be overlaborious, it is worthwhile quickly summarizing the features of each of these since the experiments are very early forerunners of what today we would call back-projected transverse axial tomography. If Takahashi had had a means of digitizing the projections in his sinogram and a computer to back-project, he could have been doing unfiltered CT. His work is important on three counts. Firstly, it was one of the first extensive investigations of the relationship between projections and cross-sections (see also chapter 4). Secondly, it generated clinical cross-sectional data which although blurred (being an unfiltered back-projection) served at the time for improved diagnosis. Thirdly, like much else here reported, these experiments were often 'rediscovered' and reperformed decades later.

2.4.1.1 Cross-section radiography as performed by Takahashi. (1) The first method he developed in 1947, he called discontinuous rotatography. The X-rays were collimated to a narrow fan, passed through a slit only 1 cm wide and irradiated the patient at discrete 20° intervals. Between each irradiation the film was advanced in its holder and when developed showed a series of strip projections. The film was then set up as shown in figure 2.10 and, for each strip, straight lines were drawn in pencil from important details in the strip (including the external outline) to a point representing the X-ray source in the same geometry as for the data capture. Each time the film advanced a strip, the paper recording the lines was rotated by an amount corresponding to the rotation of the apparatus

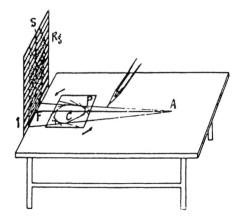

Figure 2.10 Takahashi's apparatus for reconstructing the outline of an object from a discontinuous rotatogram (discontinuous sinogram). A represents the location of the tube focus at radiography, C is a point representing the rotation centre, Rg is the discontinuous rotatogram (which is incrementally translated vertically) and P is a sheet of paper (which rotates in synchrony). (From Takahashi (1957).)

about the patient. This resulted in a line diagram, comprising 'curves' created by merging tangents, of the corresponding features in the irradiated section and by this method the outline of the lungs, bones and external contour were visualized. Takahashi developed a particularly nice display method (figure 2.11). This was needless to say a very time-consuming job being done entirely by hand and Takahashi also noted that an element of subjectivity entered in that the artist had to interpret the films by eye. However, it served the purpose. Note also that no cross-sectional *image* was made, simply a line drawing, and that this was itself somewhat discontinuous for a finite number of exposure orientations.

(2) The second method (developed in 1946) he called continuous rotatography. This was like method (I) except that the equipment rotated continuously rather than in discrete steps, thus giving a smooth sinogram which could be used in much the same way to produce outlines.

(3) The third method was continuous cross-sectional radiography (original method) (first used in 1948). This attempted to overcome the tedium of sketching from projections and the corresponding subjectivity. It was also the first method to generate an image rather than a line drawing. He utilized an optical back-projection method (figures 2.12 and 2.13). A light L_1 was shone through a lens L_2 through a strip on the film F and a slit S to irradiate at shallow incidence another film or plate P on a turntable T. The sinogram film F was translated in synchrony with the rotation of the turntable. The method is a very crude optical

130

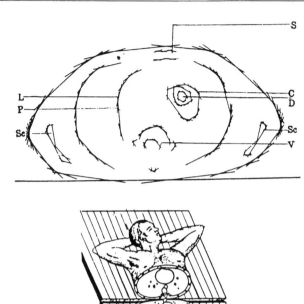

Figure 2.11 An example of the cross-section built up from discontinuous rotatography (see figure 2.10 for the method of generating this). L represents the lung, P the pleura, Sc the scapula, V the spine, C the cavity outer wall, D the cavity inner wall and S the sternum. Below is a particularly pleasing diagrammatic representation. (From Takahashi (1957).)

Figure 2.12 The continuous cross-sectional rotatograph. See also figure 2.13 for the schematic diagram and text for explanation. (From Takahashi (1957).)

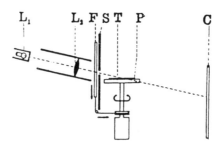

Figure 2.13 A schematic diagram of figure 2.12 showing the continuous cross-sectional radiography method. See text for explanation. (From Takahashi (1957).)

back-projection but worked after a fashion. The light was organized to fall at 7° to the plate but even so contrast was poor. There was of course no means of filtering the $1/r$ blurring function.

(4) Direct cross-sectional radiography (original method) (in 1949) was an attempt to cut out the intermediate step of forming a sinogram on film requiring subsequent optical back-projection. If a film were placed on a second table which rotated in synchrony with the patient as a narrow fan

Figure 2.14 The direct cross-section radiographic apparatus of Takahashi in action. See figure 2.15 for the schematic diagram and text for description. (From Takahashi (1957).)

irradiated a section, then an image of that section could be obtained on the film. Takahashi observed that the contrast would in this case be very poor in view of the grazing incidence of the X-rays. He hit on a novel solution (figures 2.14 and 2.15). Instead of having the film on a horizontal turntable, he arranged for the film turntable to be vertical so that the X-rays fell at normal incidence on the film. To achieve the 'back-projection' he arranged for the film to be translated across the exiting fan beam for each (discrete 20°) orientation of the patient. This achieved the desired improved contrast but it was very time consuming and associated with a large radiation dose.

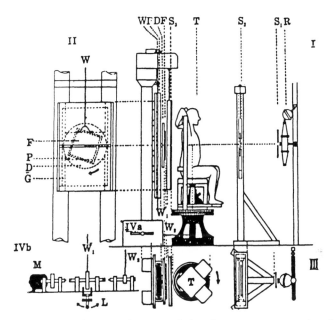

Figure 2.15 A schematic diagram of the direct cross-sectional radiographic apparatus shown in figure 2.14. R is the X-ray tube, T is the rotating patient, F is the film, D is the rotating disc, G is a groove, and S_1, S_2 and S_3 are slits. For each orientation of the patient relative to the detector and source, the film makes a complete translation to carry out a back-projection. (From Takahashi (1957).)

(5) Takahashi's fifth method, called rotatory cross-section radiography, overcame these limitations and was the closest analogy to what had been done by Watson and Vallebona independently. The patient and detector were once again on turntables with the same orientation (for example both horizontal rotations) and the narrow fan of X-rays was inclined at a shallow angle—he eventually settled on 5° to the detector. The patient

and detector were rotated in synchrony and acceptable contrast resulted. The exposure time was shortened with advantages in dose and reduced tedium. This came to be the method most commonly employed in Japan and was also used to perform inclined and curved-plane tomography in what would appear to be identical ways with those patented by Watson. He wrote, 'In Japan, Rotatory Cross-Section Radiography differs from that in other countries in that it did not evolve from tomography but developed step by step from Rotatography, so that the apparatus and technique used in Japan can be said to be characteristic of this country.' Takahashi also wrote, 'It should be pointed out here that hitherto in the living body it has not been possible to take the radiogram of the cross-section.' This was of course not so.

(6) In 1953, Takahashi produced improved versions of both the direct and the indirect methods of cross-sectional imaging. By direct is meant a method which does not require an intermediate reconstruction by optics and vice versa. He recognized that it was not strictly correct for the back-projection to be performed in parallel geometry when the X-ray beam was a narrow *fan*. To overcome this, he angled the detector, a method which has been used several decades later (see, for example, Edholm *et al* 1978). In figures 2.16 and 2.17 are shown the direct cross-sectional radiography

Figure 2.16 The direct cross-section radiograph (improved method). See figure 2.17 for the schematic diagram. (From Takahashi (1957).)

(improved method). The film detector is once again on a vertical rotating platter—as in method (4)—but, in translating across the beam at each orientation of the patient, the support for the platter travels along not a vertical plane but a plane inclined at 45°. Thus the width of the back-projected data varies with position on the detector.

(7) Discontinuous cross-section radiography (indirect method) is the improved optical version with this correction (figure 2.18). We can easily

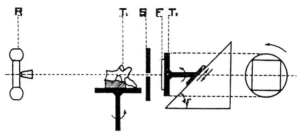

Figure 2.17 A schematic diagram of the direct cross-section radiograph (improved method). See text for explanation. R is the source, T_1 is the sample rotation table, T_2 is the second rotation table, S is the slit and F is the film. As the sample is irradiated, it rotates slowly whilst the film detector translates up the slope once for each position of the sample table. As the sample table rotates, so does the film carriage. (From Takahashi (1957).)

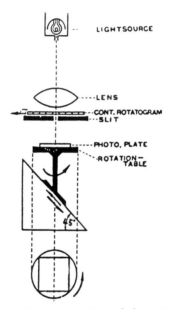

Figure 2.18 A schematic representation of discontinuous cross-section radiography (indirect method). The continuous rotatogram (sinogram) is translated slowly across the face of a slit. At the same time (and in synchrony) a film is rotated. This film translates up the slope once for each position of the sinogram. (From Takahashi (1957).)

relate this to method (3) above. The continuous rotatogram (sinogram) is translated across a slit irradiated by light as the film or plate detector rotates below at normal incidence. During the process the platter supporting the film slides in synchrony down a 45° slope, which produces the fan beam back-projection.

135

Despite the improvements in sharpness, it would appear that the tedium of these last two methods was not overcome and method (5) emerged as the most popular. By coincidence this was the method being used independently in Europe by Watson, Vallebona and others. All Takahashi's methods of course produced blurred back-projections, ignoring the need to filter the sinogram.

2.4.1.2 Solidography. The second aspect of Takahashi's work which was pioneering use of cross-sectional information was what he called solidography. This was also achievable on the Universal Rotatograph. The principle was very simple. He recognized that, when X-rays were shone at a steep angle on a rotating body, multiple sections of that body could be imaged if multiple films were stacked on the rotating film table. These images would be simple magnified versions of the true objects. Takahashi arranged for a sandwich of 120 films to be used to image a phantom. The film block when unwound and developed could be laid out to show the sections. By cutting out the regions of interest from each film and placing them on top of each other he constructed a model of the original object in three dimensions. He arranged a neat method of obtaining closely spaced sections *in vivo* without too much X-ray absorption. He placed films in a special holder at 3 cm separation, using 18 in all in a cylindrical container. Above these he placed a further 18 films. After the first set had been exposed via a complete rotation, the second set were put in place by the cylinder dropping 20 cm and another rotation was made. Following this a third set of 3 cm spaced films was exposed by dropping the film cylinder a further 21 cm (figure 2.19). In this way he obtained sections spaced at 1 cm intervals. Moreover, Takahashi constructed a photocell-driven milling machine whereby a plaster cutter was driven from the output of a scanning photocell viewing each film in turn and cutting the appropriate plaster slice. When all these slices were placed above each other, a faithful solid model of the imaged organ was obtained. By arranging for the scanning actions of the photocell and the cutter to be demagnified in the same ratio as the original magnification factor at imaging time, the models turned out to be life size. This was indeed an adventurous and pioneering development in that only today we are just beginning to witness similar developments using CT data and these machines are by no means common. Further perspective on the work of Takahashi has been provided by Gebauer (1956).

The idea of using section images to create solid objects can, however, be traced to the doctoral thesis of Ziedses des Plantes (1934a). He proposed using a series of blurred tomograms to perform what he called 'plastic reconstruction' after the terminology used in microscopy. The outlines of structures were transferred to wax plates of the same thickness as the tomogram, cut out and joined together to create solid models.

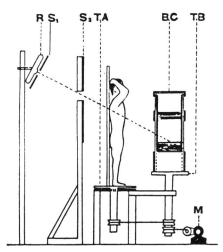

Figure 2.19 Takahashi's apparatus for solidography. The cylinder cassette is at BC, R is the X-ray tube, and TA and TB are rotating (coupled) tables operated by motor M. (From Takahashi (1957).)

The use of axial transverse tomography in radiation treatment planning became popular. Takahashi and Matsuda (1960) described their application of the Universal Rotatograph for planning rotation therapy for organs in the thorax, emphasizing the importance of the patient lying supine on a flat table and in the treatment position. They obtained sections some 0.5 mm thick using a 20° grazing angle of incidence with the film surface and used the data for planning cobalt therapy. The techniques by which they invoked axial tomography for the assistance of radiotherapy planning were covered in depth by Takahashi (1965). The subject is taken up again in § 3.1.

Axial transverse tomography was refined by Di Chiro (1963) for application to encephalography. There are diagrams in this report which show that he envisaged the same concept as Takahashi for rotating a source and a detector, rotating on its axis, about a stationary patient.

Shinji Takahashi was born in Fukushima, Japan, on 28 January 1912. He graduated from Tohoku University School of Medicine in March 1937 where he became an Instructor in the Department of Radiology. From 1954 he was Professor at Nagoya University, moving in 1974 to become Professor and Vice President of Hamamatsu University School of Medicine. He became President at the Aichi Cancer Centre in 1980. He received the Japan Academy Award in 1977, the Order of Cultural Merits in 1984 and the Sweden Royal Academy Gold Medal Prize in 1985.

Throughout his life he was a pioneer in both research and teaching, opening many courses in radiology. For a large part of his life he suffered

137

many diseases, phthisis, heart problems, stomach cancer and hepatic diseases and, as he did not expect to live long because of his poor health, he made it a rule to write a paper immediately after he finished each piece of research to set milestones in his life. As a result the milestones increased much more than he expected and in this sense his weak constitution contributed to his academic life (Takahashi, obituary 1985, Sakuma 1989). The day after he resigned his position as President of the Aichi Cancer Centre, he was titled President Emeritus at the Institute. On the following day (2 April 1985) he died.

2.5 The midcentury state of the art

The decade or so reviewed in this chapter were important years. At the outset the period was marked by a flowering of the techniques of tomography which can be traced back to Bocage. There were a few but not many commercial pieces of equipment and feeling was that these were somewhat expensive; a number of workers were developing home-made equipment as a result. The end of the 1930s saw the beginnings of interest in the UK, taking note of and adapting some of the developments in Europe. The first atlases of tomography appeared in this period and the feeling was growing that tomography was of real clinical value. There was much comment on the relation of section-imaging techniques to stereo imaging and opinions varied on the real utility of the latter. Many writers were, however, still to be convinced that tomography was unequivocally useful and it was far from artefact free. The mood appears to have been one of cautious optimism.

The most significant statement from this period was that transaxial section imaging was beginning to rival and possibly to overtake longitudinal-section techniques. The figure of Watson stands out as a pioneer of this method and there were several significant contemporary parallel developments in Europe. It might be fair to say that, whilst section imaging was initially European (non-UK) pioneered, the technique which came to be most widely used originated in the UK. It was not until the 1950s that the true perspective on the somewhat similar work in Japan by Takahashi was truly appreciated in the West. This work, however, started in the mid-1940s in ignorance of Watson's patent.

Astonishingly the patent from Frank at the outset of the period reviewed in this chapter set the scene for CT. Its timing necessitates including this work at this stage of our review but in a sense it was so ahead of its time that we shall reserve taking up this aspect of our story further until part 2.

In 1949 the pioneering radiologist A E Barclay (1949) wrote, 'Radiology has not come to the end of its expansion, and in spite of all the advances of fifty years the vista of progress still stretches out in front of us. In the fascinating country already charted and signposted there are still many by-paths to be explored even by those who now travel by motor car where the pioneers had to blaze the exciting trails before the paths were made. And for those spirits that are more adventurous there are still uncharted lands over the horizon, lands that they will open up and through which one day the roads will run.' One such land of which he could have had no thoughts is the land of CT.

Table 2.1 Some of the principal landmarks in classical tomography: the years of consolidation, circa 1940 onwards.

Year	Worker; invention
1936	Naud; doctoral thesis
1936	Massiot; the commercial Planigraph
1937	Massiot; the Biotome of Dr Bocage first exhibited (October)
1937	Colyer; work with a Grossmann Tomograph, and home-made tomography in the vertical position
1938	Bocage; first paper after his 1921—2 patent describing the Biotome
1938	Watson; first commercial Sectograph
1938	Taylor; home-made tomograph
1938	Keleket Company (USA); the Kele—Kieffer
1939—40	Watson; US patent
1939	Chausse; multiple firing of beam during tomography
1939	Jones and Bradley Bowron; Bozzetti-type vertical tomograph constructed in India from scrap materials
1939	Bodle; tomograph from scrap materials
1940	Viswanathan and Kesavaswamy; two tomographic devices in India based on rod and pulley mechanisms respectively
1940	McDougall; atlas of tomography
1942	Olsson; 'improved Mayer method' of use of motion in imaging
1943	Goldman; home-made (War-time) equipment for planigraphy
1944	Baylin; clinical work
1944	Didiee; experiments
1945	Takahashi; 'rotation radiography' including axial transverse tomography, Grossmann-type tomography, and experiments, precursor to CT and solidography
1945	Brock; home-made (War-time) equipment for planigraphy
1945	Robin; home-made (War-time) equipment for planigraphy
1946	Weinbren; atlas of tomography

Table 2.1 *(continued)*

Year	Worker; invention
1947	Frain and Lacroix; apparatus for axial transverse tomography
1948	Vallebona; apparatus for axial transverse tomography
1949	Gebauer; apparatus for axial transverse tomography
1949	Paatero; pantomography
1949	Sans and Porcher; prototype Polytome
1950	Watson; multiple-cassette method
1950	Watson; Sectograph in use with Stephenson
1950	Janker; axial transverse tomography in recumbent position
1950–4	Paatero; patent for pantomography
1951	Massiot; Radiotome
1951	Sans and Porcher; the Polytome unveiled in Brussels
1953	Watson; layer fluoroscopy experiments
1954	Giacobini and Manzi; axial transverse tomography
1954–63	Verse; patent for tomographic fluoroscopy
1955	Griesbach and Kemper; review of commercial equipment
1956	Gebauer; short history of tomography
1956	Lasser and Nowak; multisection imaging
1959	First commercial Orthopantomograph
1962	ICRU; official adoption of the term 'tomography' for all body-section imaging
1962	Watson; axial transverse tomoscopy using television and rotating prisms
1964	Farr; book on section imaging at Manchester
1964	Bruwer; *Classic Descriptions in Diagnostic Röntgenology* appears with short history of tomography by Bricker
1966–70	Richards; patent for electronic seriescopy
1966	Westra; book on zonography
1969	Takahashi; atlas of transverse axial tomography
1970–3	Richards; improved seriescopy (patent)
1973	Ziedses des Plantes; selected works published in honour of his seventieth birthday (including English translation of his 1934 thesis)
1973	Westra; short history of tomography
1974	Massiot; reflective short history of tomography
1976	Littleton; encyclopaedic work on tomography published

Chapter 3

Circa 1950 and Beyond

3.1 The blossoming of transverse axial tomography

At the Sixth International Congress of Radiology, held in London in 1950, there were several papers on body-section imaging. Vallebona (1950a) was still very active in this field. His paper reported his invention of equipment for transverse axial tomography in 1947. We recall the work in 1937 of Watson. There was no doubt much discussion since Watson (1950, 1951) presented his improvements to his pre-War equipment, enabling multisection imaging, and Stephenson (1950a) was also presenting the results from transverse axial equipment in London. The presentation by Watson was the first ever to be given by a radiographer at any international congress of radiology. This work on multisection imaging was described more fully by Stephenson (1950b) and by Watson (1953a). Schonander (1953) noted contemporary developments including the 'Universal Planigraph' shown at the exhibition accompanying the London congress. Watson's development of multisection radiography included overcoming the problems of ensuring adequate sensitivity of all the layered detectors. At the time, only Ilford Limited were producing salt-intensifying screens in the UK although these were supplied to other companies and renamed (Mercer 1988). As the screen unit was not a mass mechanized process, it was relatively easy to produce special coatings for trial purposes.

Stephenson (1950a) reviewed only those historical developments that specifically related to transverse-section imaging, noting the work of both Watson and Vallebona. He recalled that Watson's second-generation equipment was commercially produced by the Medical Supply Association Ltd in 1939. In this equipment the turntable was driven via a four-speed gearbox (figure 3.1) and typically top gear, giving a complete revolution in 3 s was engaged. Exposures as fast as 1 s were made but this was 'very disconcerting for the patient' (Watson 1962) and there was no

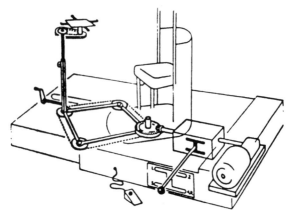

Figure 3.1 The method of rotating the patient and the detector in the Mark 2 equipment of Watson for transverse axial tomography (the Sectograph). Notice the use of the car gearbox to vary the speed! (From Stephenson (1950a).)

real advantage in the fast times. The angle of the tube was 25–35° to the horizontal, giving a slice thickness of some 1.5 mm. Stephenson noted that many other workers had developed equipment for transverse-section tomography without being aware of the work of Watson. He cited Kieffer (1938), de Abreu (1944), Herdner (1948a,b), Frain and Lacroix (1947a, 1948) and Vallebona (1948). Gebauer (1949) also produced equipment and a commercial version with Siemens. Stephenson was using Watson's machine and acknowledged his help in setting it up at the (then) Royal Cancer Hospital (see §2.4). Most of his paper concerned clinical cases.

Vallebona (1950b) summarized his own work in body-section imaging and placed it in context with that of other workers. His experiments in transverse-section imaging were, he said, 'fruitful' in 1947 for *in vivo* work. He laid claim to having invented the title 'axial transverse laminography'. This paper shows a commercial version of his equipment, manufactured by the Zuder Company of Genoa. In this, imaging was through a full 360° and took between a half and several seconds. The equipment was used for imaging the chest, mediastinum and base of the skull. Again surprisingly this paper makes no mention of the work of Watson pre-War. A great deal of clinical work was performed in Genoa under Vallebona. For example images of the mediastinum are shown by Buzzi (1950) (see also the references therein).

The techniques of Twining and Grossmann had become routinely established in the 1950s. McInnes (1954) reviewed tomography for a radiographer's journal. It was well known that the slice thickness depended inversely on the size of arc of travel and hence, for a fixed linear travel, on the focus-to-film distance. The smaller the focus-to-film distance, the greater is the angle and thus the smaller is the slice thickness. Contemporary developments had, however, enabled the slice thickness to become independent of the focus-to-film distance. It was also noted that a combined Twining–Grossmann method had become popular whereby the tube and detector travelled in constant-separation parallel planes, but the angle of the tube rotated whilst moving in this way, thus enabling the use of tighter beams. There was some emphasis on dose reduction, it being noted that tomography should be performed for a limited number of cuts at pre-determined locations. Tomography was proving useful for the odontoid process, and for the cervicothoracic region—specifically for imaging spina bifida. Multisection imaging with book cassettes was routine. Indeed a number of papers reported developing schemes for multiple-section imaging (see, for example, Lasser and Nowak 1956) which are reminiscent of the suggestions in Watson's patent.

Experience showed that great care was necessary in setting up equipment for transverse axial tomography. In order to obtain useful tomograms, it was vital that the mechanisms for rotating film and patient were precisely synchronized and that the position of the focal spot of the source was within ½ mm of the common plane of the two axes of rotation. It was felt in Manchester that sufficient difficulty had been found there in setting up a 'Zuder Assis-Strator' that a textbook was required to guide others (Farr *et al* 1964). The Manchester equipment was considerably modified including provision of a collimator to reduce scatter. Interestingly the equipment is still in almost daily use (Massey 1988). This monograph also contained an extensive atlas of tomograms produced by this technique which might be contrasted with the atlas produced by McDougall some two decades earlier.

The situation in France with regard to the use of transverse axial tomography was reviewed by Bonte *et al* (1955). The method appears to have been fairly popular. By the early 1960s, Watson (1962) was able to record, 'In Great Britain interest in transverse tomography has been rather lukewarm, but it is gratifying to see that it is now increasing somewhat, 25 years after its inception.' Watson also proposed that intermittent exposures might be advantageous in transaxial tomography in view of the fact that, when the beam entered the patient laterally, very little contribution was made to the exposure and most of the dose was absorbed by the patient. He attempted several arrangements to cut off the radiation for sectors up to 45° during the rotation.

Watson (1957, 1962) also proposed apparatus for transaxial tomoscopy (figure 3.2). Instead of the film as detector a fluorescent screen was placed on the rotating turntable as the detector. This was viewed directly above by a television camera (preferably with an image-intensifying arrangement in view of the very weak fluorescent image). Below the television

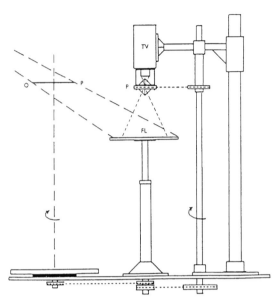

Figure 3.2 The apparatus for axial transverse tomoscopy proposed by Watson. The plane OP is the plane in focus in the rotating patient (on the left turntable); FL is a fluoroscopic screen (which need not rotate but could do so synchronous with the patient rotation). A television camera TV views FL through the rotating prisms P which are coupled via a 2:1 gearing to the patient turntable. (From Watson (1962).)

camera lies a pair of prisms joined along their hypoteneuse and rotating at half the speed of the detector turntable and in the same sense. Under these conditions a stationary image is recorded of just one plane in the patient. A little thought shows that there is no necessity for the detector to rotate at all provided that the prisms rotate at half the angular velocity of the patient turntable. The idea has a lot in common with the rotating-mirror arrangement of Ponthus and Malvoisin (see §1.2.8) for plan-igraphy although Watson himself did not draw this connection. The use of television is significant and, as we shall see in §3.5, Watson and others were also using television to perform electronic longitudinal-section seriescopy.

Watson (1956) put forward a number of innovations including per-forming transverse axial microtomography with section thicknesses of some 20 μm, slit pantomography (see also §3.2) and Grossmann-type section imaging on a ^{60}Co therapy machine.

In 1969, Shinji Takahashi (1969) produced an atlas of axial transverse tomography and its clinical applications. This volume is remarkable not just because it is the most complete coverage of the whole body with images of transaxial slices but also because it embodies concepts which have directly been taken over into CT and which some regard as having derived from this newer modality. For each section in the body a trans-axial tomogram was coupled to the corresponding anatomical drawing and the level of cut was shown on a planar anteroposterior radiograph with a schematic of the reclining patient. This practice is precisely that which has been so effectively reused with images from X-ray CT in more recent years.

The volume only just pre-dated commercial X-ray CT (see chapter 4) and one would not expect anyone to produce an updated compendium in view of the superior quality of X-ray computed tomograms. The book is thus destined to remain a classic piece of history and became outdated with unfortunate rapidity. Nevertheless the images shown are quite start-ling and, with skilful interpretation, axial transverse tomograms were most useful. In the atlas there is a comprehensive coverage of normal anatomy followed by a treatment of selected pathological cases.

Takahashi noted that axial transverse tomography had not become as popular as he had expected and he attributed this partly to the practice of imaging the upright rather than the reclining patient. For his work he adopted the proposal from Janker and imaged the patient lying down. He also proposed a viewing convention whereby cross-sections were displayed as they would appear to a viewer at the feet of the patient (that is right shown left and vice versa; posterior to the bottom of the display). This mode of display was subsequently adopted for CT. He also suppositioned that axial tomography was hampered by the doctors' lack of familiarity with viewing anatomy in transaxial cross-section (whereas longitudinal tomograms bore a passing resemblance to conventional planar films as viewed since X-rays were discovered). His book, with its beautiful anatomical drawings went a long way towards correcting this and of course today radiologists for whom CT is a natural part of their repertoire would have no such difficulty.

What really distinguishes this atlas is how it couples diagnostic axial tomography with radiotherapy treatment planning. The utility of the tomograms arises from being taken with the patient in the same geo-metrical position as they would be at the time of treatment. 20 years on, this principle is still being stressed and its importance misunderstood by some. Others who at the time found axial tomography of less use for treatment planning were hampered by using films taken with the patient sitting upright. What is more, Takahashi showed how corresponding tomograms could be taken on the treatment machine (he called this 'beam focus radiography') for verification of confirmation of the treatment set-up. The arrangement is shown in figure 3.3. His book explains how the two axial tomograms taken on the simulator and on the treatment machine could be used to align the patient correctly. These are very modern ideas and many of us would argue that they are still not fully

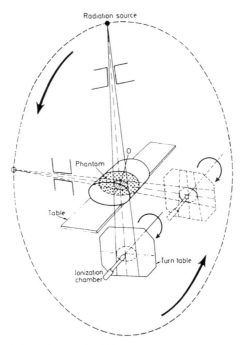

Figure 3.3 Takahashi's method for imaging cross-sections of the patient on the treatment machine with the patient in the exact same position as at simulation with axial transverse tomography. He called this 'beam focus radiography'. The centre of rotation of the film is on the line joining the X-ray source to the centre of rotation of the unit. As the X-ray source rotates anticlockwise, the cassette rotates clockwise. (This is as if the source had remained stationary and the *patient* and cassette had rotated in the same direction.) (From Takahashi (1969).)

developed. The use of X-ray CT data for treatment planning has of course become widely popular but not many centres verify set-up at the point of treatment using CT data. The name 'conformation radiotherapy' was publicly announced in 1960 meaning 'rotation arc therapy' (see also Takahashi 1965) and not as the term is used today to mean tailoring the high-dose volume to the tumour. The latter is only now about to become a feasible proposition.

Most of the early apparatus for axial transverse tomography generated slices with the patient upright; that is the slices were parallel to the floor. For example Pierquin (1961) used the Massiot–Philips apparatus known as the Radiotome, based on the ideas of Frain and Lacroix (Massiot–Philips, undated but circa 1950), to assist radiotherapy planning. The noticeable exception was the work of Takahashi. When it came to investigating the usefulness of axial transverse tomograms for radiotherapy treatment planning, it was recognized that to be effective the images had to be obtained with the patient in the treatment position, that is generally lying down. With this in mind, Kishi *et al* (1969) developed a machine, marketed by Toshiba, whereby the patient lay horizontally stationary whilst the X-ray source executed a vertical circle and the film cassette moved in the same vertical circle whilst also rotating about its own axis. The body contour and internal lung contours were reasonably well defined (figure 3.4). In a sense this was the forerunner of the philosophy

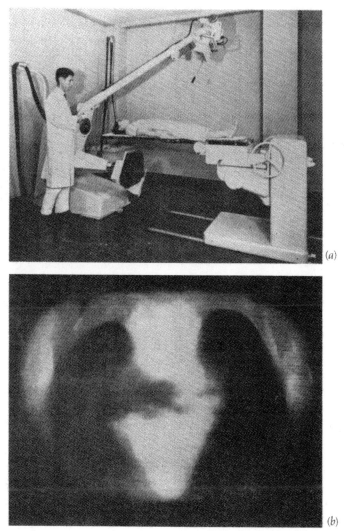

(a)

(b)

Figure 3.4 (a) The axial transverse layergraph developed by Kishi *et al* to take transaxial sections of the patient lying supine in the treatment position for radiotherapy. (b) An example of a chest cross-section taken with this apparatus (73 kV$_p$; 50 mA; 12 s). (From Kishi *et al* (1969).)

147

behind recording 'therapy CT scans' with the patient supine in the treatment position.

So we see that axial transverse tomography was poised to play the role usurped by X-ray CT and to a large extent the latter shut down in response. However, the historical precedent shows that many ideas which some trace to the origins of CT can in fact be traced back much further.

3.2 Curved-plane tomography

We recall that all forms of tomography require the source and detector to execute proportional movement about a fulcrum which defines the plane in focus. It follows fairly trivially from this principle that, if when radiographing a solid body with a flat surface the detector is placed very close to this flat surface and at the same time the fulcrum is arranged to be very close to that surface, then whilst the source executes a large movement the film will hardly move at all. The plane in focus is that passing through the fulcrum just above the film. As a good approximation to this form of tomography the film need not move at all. This is a very old principle dating back as we saw in §1.2.13 to the pioneering work of the Polish radiologist Karol Mayer (1916). The structures close to the film will then be seen in focus and overlying structure will be blurred. The same is true whatever the form of movement of the X-ray tube. Olsson (1942, 1944) developed a form of tomography whereby the X-ray tube moved normal to the detector which was placed immediately behind the body being imaged. Structures close to the film were then imaged sharply whilst those in other planes were blurred by this procedure of varying the focus-to-film distance.

This is the principle underlying the important development by Paatero (1949). He thought it would be very interesting to be able to image sharply the curved outer surface of the cranium onto a film which could then be laid out flat for inspection. He drew the analogy of making plane maps of the globe whereby oceans and land on opposite sides of the globe could be seen simultaneously. Figure 3.5 shows how he achieved this. D represents the outer curved surface of the cranium. A film K is wrapped around a semicircular portion of this surface. The film and the cranium remain stationary whilst an X-ray tube sweeps around the other semicircular portion. The X-ray beam is collimated to a narrow vertical strip. The figure shows three positions of the source FI, FII and FIII. Rays B, B_2 and B_3 are three positions of the extreme rays from this narrow fan. C is the central ray. It may be easily appreciated that whilst, for example, the source moves from position FI to FII, an image point a' close to the surface is imaged at more or less the same point a'' on the curved film by ray A from position FI and by ray B_2 from position FII whereas these rays are passing through very different body components, causing their images to be superposed as a blur. In this way it is clear that a sharp image on the film is obtained of that curved surface next to it with a superposed background of blurred shadows. When the film is unwrapped, an 'unfolded' cranial map emerges showing half of the cranium. A schematic diagram of the practical equipment is shown in figure 3.6.

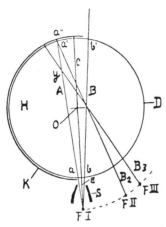

Figure 3.5 How structure close to a film wrapped around a subject is relatively unblurred when irradiated by an opposing source. This is the principle of Paatero's curved-plane tomography. For detailed explanation see text. (From Paatero (1949).)

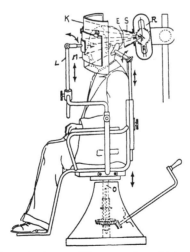

Figure 3.6 A schematic picture of the apparatus which makes it possible to radiograph half of the cranium spread like a plane. R is the X-ray tube, F is the focus, S and E define a narrow vertical swathe of beams, K is a semicylindrical or parabolic cassette, O is the axis of rotation of the chair, and L is a forehead support with a cassette holder M. The chair is rotated by hand or by an electric motor. (From Paatero (1949).)

149

The source was stationary and the film and patient rotated. Figure 3.7 shows images obtained by such a means. Paatero also invented a means of recording a whole-cranium tomogram (figure 3.8). Obviously the film cannot be wrapped right around the full circumference and instead the film is contained in a cassette holder B, with only the central part of the film exposed through a vertical slit. As the patient is rotated in front of the stationary tube, the film is translated behind this slit to achieve a sequential exposure of the circumference of the cranium. Note that in the figure (profile) the (scatter-rejecting) screen and film are placed very close to the head, somewhat remote from the mechanism B which effects the translation. In the plan view the film has been drawn close to the cog drive. The half-cranial equipment is clearly the simplest but is able to record scatter. The full-cranial equipment has a mechanism for scatter rejection but is somewhat more complicated. The work of Paatero is the forerunner of modern dental pantomography. In the opinion of Vuorinen (1959) the work of Paatero is the only slit tomographic technique of real practical importance, other methods, for example those of Heckmann and Vieten (see §§1.2.14 and 1.2.15), being somewhat impractical. The first commercial Orthopantomograph was built in 1959 by the Finnish manufacturer Lääkintäsähkö Oy and distributed by Oy Sarjavalmiste AB of Helsinki. It has been remarked by critics that, in view of the unpopularity of pantomographs, the Swedish definition of tomography in

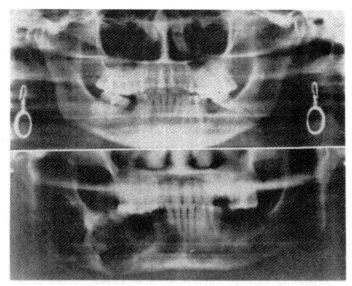

Figure 3.7 Semicranial röntgenograms produced by the method shown in figure 3.6. The top röntgenogram shows a fracture of the mandible on the left, and the bottom röntgenogram an adamantinoma in body and ramus of lower jaw on right. (From Holt *et al* (1955).)

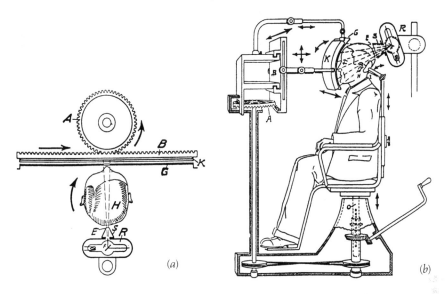

Figure 3.8 A schematic picture of the apparatus which makes it possible to radiograph the whole cranium spread like a plane. A is a cog wheel linked to the cassette holder B holding a curved cassette K. G is a lead screen to prevent the scattered radiation, R is the X-ray source, F is the focus, S is the source collimator and I is a translucent neck support. The arrows show the possible adjustments. The method is described in the text and was superior to hemicranial pantomography as performed with the equipment in figure 3.6 because of the possibility of scatter rejection. (From Paatero (1949).)

general is 'an expensive way of getting poor pictures' (Grigg 1965). The editorial comment in the *1955—56 Yearbook of Radiology*, however, ran thus: 'Paatero's technique, mentioned in the *1950 Yearbook* appears to have stood the test of time. . . . We continue to regard it as a fascinating ingenious method, the results of which are positively startling' (Holt *et al* 1955—6). Paatero (1955a) continued with the technique into the mid-1950s; indeed he also developed a method of stereopantomography in which two films were simultaneously exposed with necessary modifications to the equipment (Paatero 1955b). Paatero was a teacher of dental radiology at the University of Helsinki and lived from 1901 to 1963. Paatero patented his technique (Paatero 1950—4). It has, however, to be recognized that Bocage's patent, the very first in tomography, did mention the possibility of obtaining curved sections.

It is interesting to note the similarity between these principles and those by which certain structures could be imaged more sharply by introducing movement to superposing structure. For example antero-posterior radiographs of the superior cervical spine were improved by the patient opening and closing the mouth (Pelissier 1931, Ottonello 1930,

1935, Jacobs 1938). Ziedses des Plantes (1937) rotated the head during radiography of the odontoid process and Jönsson (1937) requested the patient to continue breathing during radiography of the sternum to blur structures which were not wanted in focus. Holly (1942) has reviewed these methods with a neat analogy: 'The radiographic result (of including movement) is somewhat analogous to a photograph of a street scene made with a small lens aperture and a long exposure, in which the images of buildings and other stationary objects are sharp while the images of moving people or vehicles are either blurred or virtually invisible.'

3.3 Watson and layer fluoroscopy

In § 1.2.5 we saw how, in the patents from Pohl, the possibility of visualizing longitudinal tomographic sections on a fluoroscopic screen had been proposed. The tomographic effect was achieved differently. Whereas with a film or plate as detector the effect is produced by ensuring the shadows of certain structures are stationary on the detector whilst others, in out-of-focus planes, move, when a fluorescent screen is used the effect is psychological. If certain structures can be rendered stationary to the eye whilst others move as time passes, the integrating eye can distinguish between them and essentially discard the moving structures in favour of the stationary ones. The idea also surfaced again with Kieffer and the work of Ponthus and Malvoisin (see § 1.2.8) also made use of this principle.

Designs for a practical implementation did not come until much later when Watson (1953b, 1957) provided effective embodiments to the idea. There was not much difficulty arranging for the X-ray source to rotate on a circular trajectory and his method is shown in figures 3.9 and 3.10. An independent shock-proof tube connected to a remote high-tension transformer was situated on the baseplate in the upper part of figure 3.10 which in turn was supported by three or more rods b having universal joints at each end. The supporting rods were inclined inwards so the tube could gyrate at any desired radius whilst the central ray always passed through a given point. This radius was controlled by a lever a which adjusted the throw of the links. The arrangement can be seen in the lower part of figure 3.9.

The ingenious part came in the manner of arranging for the image from just one plane to appear stationary. Above the fluoroscopic screen lies a mirror, inclined to the plane of the screen and rotating upon its axis in synchrony with the gyration of the X-ray source. Additionally *and most important* the plane of the mirror is tilted with respect to its own axis of rotation. This can best be seen in figure 3.11, the tilt being variable and effected by the small lever shown. Thus, as the plane mirror rotates, there

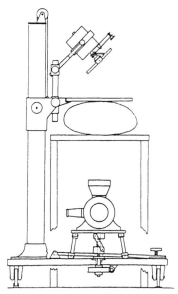

Figure 3.9 Watson's arrangement for layer fluoroscopy. The patient is represented by the ellipse above which is the flat fluoroscopic screen. Above the screen is the mirror with swashplate action (seen close up in figure 3.11) and beneath the patient lies the gyrating X-ray tube. (From Watson (1953b).)

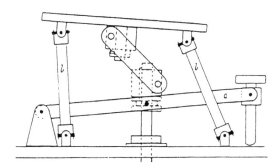

Figure 3.10 The mechanism of controlling the gyration of the X-ray tube which is situated on the baseplate supported by rods b which incline inwards and whose inclination, controlled by rod a, effects the angle of inclination of the X-ray tube to the axis of rotation and the radius of rotation. (From Watson (1953b).)

is an apparent wobble or swashplate action. With the correct angle, just one of the gyrating images on the screen appears stationary to the viewer. By changing the tilt angle of the mirror different planes can be selected. Watson (1953b) provided ray diagrams to show why this should be the case and also built an optical model of the arrangement.

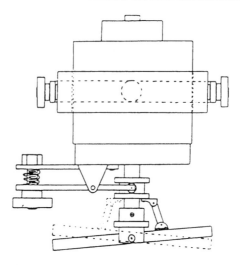

Figure 3.11 The motor-driven mirror unit. The angle of the mirror to its own axis of rotation is adjustable to select the plane in focus. Two positions of the mirror are shown corresponding to half a rotation apart. (From Watson (1953b).)

Precise synchrony with the phase of the X-ray source was also arranged. A second plane mirror viewing the first was used simply to correct for lateral inversion.

Watson obviously had several tries at this. He wrote, 'From time to time (I have) attempted a more practicable mechanical solution to this problem (than Pohl).' He first tried to effect the pseudostationarity of the selected image by a rotating refractor but this met with less success. In time he expected that mechanical methods would be replaced by electronic means as electron optics developed. In chapter 6 we shall see that not dissimilar ideas underlay some methods of tomography in nuclear medicine (such as rotating-collimator tomography) which were indeed achieved electronically. Watson concluded, 'It is hoped that an X-ray machine of this kind will soon be in course of construction.' The methods by which Watson achieved what eluded Pohl were all patented (Watson 1947–8, 1951–4).

3.4 Commercial tomography

Before the War there were very few commercial pieces of equipment for performing body-section imaging. The first commercial unit was the Planigraph (in 1936) from Massiot et Cie who also made (in 1937) the Biotome of Dr Bocage. In America the Kele–Kieffer appeared (in 1938).

The commercial position changed during the 1950s and 1960s with many commercial products. These have been reviewed in detail by Littleton (1973, 1976), by Littleton and Raventos (1965) and by Griesbach and Kemper (1955). The latter (in German) is very detailed and contains a wealth of information on commercially manufactured equipment. These books and papers also contain photographs of many of the commercial tomographic sets. The giant history of radiology by Grigg (1965) is also a mine of information on commercial companies and their history including many details of tomography. Table 3.1 summarizes some of the commercial tomographic equipment. These machines differed not only in the type of blurring which they offered but also in the distance through which the tube moved and its angle to the axis of rotation which in turn led to a range of slice thicknesses. Provision was also made for varying the exposure time. In practice, linear tomography was generally not preferred when pluridirectional movements were available on the same machine. More attention was also being paid to test objects for tomography (see, for example, Le Clerk *et al* 1967−70). Notice in the table the fanciful names which were invented. One wonders what the patients made of these.

Littleton (1976) published a monumental atlas of tomographic imaging and prefaced this with a miniature historical review. This, like many of the mini histories which have appeared from time to time, recounted the multiplicity of reinventions of classical tomography. However, it also provided a frank comment that tomography was nothing like so popular in the USA as in Europe. During the period 1940−50, 758 articles in the foreign literature were devoted to clinical tomography whereas only

Table 3.1 Commercially produced equipment for tomography.

Trade name of equipment	Firm	Features
Ascistrato	Gilardoni, Como	Horizontal body tomography
Assis-Strator	Zuder, Genoa	Axial transverse stratigraphy
Baltom X	Balteau, Liege	Linear tomography
Biotome	Massiot et Cie	Experimental implementation of Bocage's principles
Buckystrat	Medicor, Budapest	Horizontal tomography
Craniotomo	Barazzetti, Monza	
Danatom	DRT, Copenhagen	Multisection tomography
DG 101	Gera, DDR	Universal planigraph

Table 3.1 *(continued)*

Trade name of equipment	Firm	Features
DG 102	Gera, DDR	Multisection linear tomography
Eurostrator (Gyratome)	Zuder, Genoa	Spiral, circular and double-linear trajectory
Goniotomo	Barazzetti, Monza	Horizontal and vertical tomography
Horizontal Planigraph	Massiot et Cie	Implementation of Ziedses des Plantes' planigraphy
Laminograph	Keleket	Implementation of Kieffer's laminography
Layergraph	Sanyei, Osaka	Horizontal laminography
Layergraph	Toshiba, Tokyo	Circular and hypocycloidal
Layergraph type D	Toshiba, Tokyo	Horizontal multilayer
Maxitome	General Electric Medical, Liège	Linear and pseudosinusoidal trajectory
Mimer	Elema–Schonander, Stockholm	Linear and circular trajectory
Multiplanigraph	Siemens–Reiniger–Werke, Erlangen	Linear, circular or elliptical trajectory
Neotomo	Barazzetti, Monza	Horizontal tomography
Ordograph	General Electric	
Orthodiagraph	Siemens–Reineger, Erlangen	Dental pantomography
Orthopantomo-graph	Palomex, Finland	Semicircular pantomography
Oscillostrator	Companie Générale de Radiologie, Paris	Horizontal, linear
Panorex	S S White, Philadelphia	Dental pantomography
Panorex	XRM, Long Island	Dental orthopantomography
Panoramic	General Electric Company	Dental pantomography
Pantomix	Companie Générale de Radiologie, Paris	Axial transverse tomography
Pantomix	Companie Générale de Radiologie, Paris	Transverse body section (de Vulpian)
Planigraph	Siemens–Reineger	Tomography (patient standing)
Planitom	Chirana, Prague	Multiplanar horizontal

Table 3.1 *(continued)*

Trade name of equipment	Firm	Features
Polystratix	CGR Gencray, Monza	Circular and elliptical trajectory
Polytome	Massiot–Philips, Paris	Linear, circular, elliptical or hypocycloid trajectory (after ideas by Sans and Porcher)
Plurigraph	Picker Gmbh, Estelkamp	Elliptical and circular trajectory
Pluristrator	Zuder, Genoa	Linear and circular trajectory
Radiotome	Massiot–Philips, Paris	Axial transverse tomography (after Frain and Lacroix)
Rotagraph	Watson	Orthopantomography
Sectograph	Medical Supply Association	Implementation of Watson's differential radiography
Stratigraph	Meschia	Implementation of Vallebona's stratigraphy
Stratix	Barazzetti, Monza	Pluridirectional body sections
Stratix	Companie Générale de Radiologie, Paris	Implementation of Bozzetti's stratigraphy
Stratix	Companie Générale de Radiologie, Paris	Vertical stratigraphy
Stratagraph	Westinghouse	Planigraphy
Stratomatic	Companie Générale de Radiologie, Paris	Linear, circular and trispiral trajectory
Telestrator	Zuder, Genoa	Stratigraphy
Tomofluorograph	Ovoshnikov, USSR	Tomofluorography
Tomograph	Picker	Linear tomography
Tomograph	Sanitas	Implementation of Grossmann's tomography
Tomoscop	Nowikow	Russian tomography
Transversal Planigraph	Siemens	Transaxial tomography after Gebauer
Transversotome	Electrik, Linz	Axial transverse tomography
Trocoradio-stratigrafo	Rangoni–Puricelli, Bologna	Linear tomography
Ultragraph	Isotopan, Rome	Circular trajectory
Universal planigraph	Siemens	Horizontal and vertical tomography

107 articles appeared from American workers. During 1950–60 the corresponding figures were 2255 and 150. A survey of radiographic equipment usage in the USA conducted by the Radiological Society of North America in 1962 showed that 65% of US radiologists had a tomographic device, all of them with linear obscuring movements. Of these, however, only 6% used them for 1% of examinations. The lethargy was thought to be in view of the well recognized limitations of linear obscuring movements and the situation improved when pluridirectional machines became more common.

The modern era of tomography was characterized by the pluridirectional units. The first of these was the Polytome, developed by Jean Massiot from plans patented by Sans and Porcher (Tobb 1950). The prototype was exhibited in Brussels in 1951 following good experience with a pre-production machine at the Saltpêtrière Hospital. The prototype went to the Electroradiology service of l'Hôpital de la Pitié in Paris. Two features of some importance characterized the Polytome. First the patient could be sectioned in either the vertical or the lying-down position, the former being most useful for pulmonary imaging and the latter for imaging the head. Secondly a number of pluridirectional movements could be obtained by an orthogonal arrangement of two motors which drove the linked source and film. With one motor switched off, a linear trajectory was obtained; elliptical trajectories resulted from sinusoidal movements in the two directions and for extremely accurate work with bony structures a complicated trefoil-with-four-leaves trajectory was used. By 1960 there

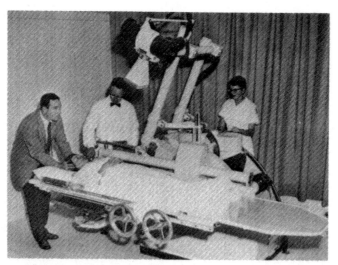

Figure 3.12 The first commercial pluridirectional device installed in the USA, a Polytome. Left to right C L Rumbaugh, J T Littleton and V Wilcox. (From Littleton (1976) *Tomography: Physical Principles and Clinical Applications* The Williams and Wilkins Company. Reproduced with permission.)

were some 100 Polytome units in use in western Europe. The first machine (figure 3.12) was set up in the USA by Littleton in August 1960 imported by the Schick X-ray Corporation, Chicago (Littleton *et al* 1963). The second Polytome in the USA went to the University of Chicago in 1961 and gradually through the 1960s the emphasis in the USA shifted to the use of this and similar devices (Littleton 1976). The sheer size of Littleton's compendium shows how important tomography had become after 1960. The work also contains a section devoted to veterinary tomography. Some of the earliest practitioners of this science went through the same phases as their counterparts working with human patients of developing in-house hand-made attachments for tomography before moving on to use commercially available equipment.

Tomography also found a role in the X-ray examination of stamps and paintings (Pease 1946, Bridgeman 1965). According to Pease (1946), about 5% of the radiographs taken at the Metropolitan Museum's laboratories benefited from tomographic imaging. This enabled the image to represent more properly the varying densities of the paints used (and thus possibly to show detail subsequently obliterated to the naked eye view) and to 'remove' the unwanted and confusing shadows of the panels on which the paintings were mounted. Such might include the images of nails, heavy cradle members, putty, sealing wax, white lead paint and so on. The method used was attributed to F I G Rawlins at London's National Gallery. This was in fact not so much a tomographic technique as an application of Mayer's principle (see §1.2.13). The X-ray film was placed in contact with the paint surface and an image recorded as the X-ray source traversed an arc (figure 3.13). In practice this was usually

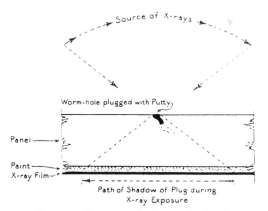

Figure 3.13 Schematic diagram of the method of traversed focus radiography for paintings. During the exposure the density pattern of the paint layer in contact with the X-ray film is recorded sharply, regardless of the angle of the rays. The shadow of an obstruction not in contact with the film moves during the exposure and is either blurred or lost altogether. (From Pease (1946).)

done by pivoting the panel–painting–film about its centre in a stationary X-ray field. The image of the paint remained stationary on the film whilst the shadows of structures, such as the worm hole shown, more distant from the film, were blurred. The method was called 'traversed-focus' radiography. In this paper the effect was demonstrated for *The Nativity* by Fiorenzo di Lorenzo.

3.5 Late development in laminography

We have seen how Watson's work with a fluoroscopic screen enabled layered fluoroscopy to be performed. His work, cross-linked back to those early ideas from Pohl, still used mechanical methods. However, he astutely recognized that soon electronics would be able to achieve more conveniently the tasks of these mechanical methods. Richards (1966–70, 1970–3) patented techniques for multiple-plane laminography. These relate to the methods which were described in § 1.3 for analogue multi-section laminography or seriescopy. We recall that the origin of this idea was with Ziedses des Plantes with considerable work having been done also by Cottenot. Largely this flourished in the mid-1930s. Here was the old idea making a reappearance.

In the first patent, Richards explained that good laminography depended on the number of films which were superposed (with lateral shifts) to render one plane in focus. Because it was clearly impractical to use large numbers of films he proposed to capture the planar images digitally using a fluoroscopic screen viewed by a television camera, for the many orientations of the X-ray source and detector relative to the patient. Then, by *electronically* reregistering these data subsequently, various planes were brought into focus. We shall see in chapter 6 how very similar ideas were used in connection with emission laminography.

In the second patent, Richards returned to the idea of using film as detector and restated that, if several distinct films were exposed with the source at a series of different positions, then, by selectively registering these films in different ways, selected layers in the patient could be brought into sharp relief with structures out of these planes largely blurred. The device used to perform this was the seriescope. What was new in Richards' patent was a rather clever idea for making the registration simple. He suggested that the films should all be cut to the same height. The other two sides of the films were to be cut so as to form a parallelogram. The width of each film at half-height was the same but the *inclination* of the two sides of the parallelogram in width was varied in proportion to the linear travel of the source. Figure 3.14 makes this clear. At one particular location of the source, that half-way along its travel, the film was cut square. Either side of this, films were cut at progressively

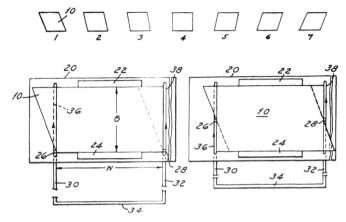

Figure 3.14 The Seriescope patented by Richards in the 1970s. Films were exposed with the X-ray source in various positions. The sides of the films were cut to form parallelograms of varying slope as shown in the upper part of the figure. The films were then placed on a viewing box between guide rails 22 and 24 and thus constrained to move only horizontally. Pins 26 and 28 made vertical traverses in slots 36 and 38; hence the amount of lateral movement in the Seriescope depended on the angle of cut of each film. The differential film movement gave the impression of selective focusing at different depths. (From Richards (1970–3).)

steeper angles. When these films were all superposed in the seriescope, selected layers were brought into focus simply by moving two pegs along the angled sides to control the relative sliding of the sets of films. Since this will cause the films to move at different speeds in the orthogonal direction (that of the source movement), the manner of focusing is obvious.

Verse (1954–63) patented a method of tomographic fluoroscopy whereby an X-ray source and an image intensifier could be moved in circles to perform laminography. In his patent he provided diagrams showing how several of the accepted tomographic movements could be combined with the use of a luminoscope for this purpose, including linear movement, circular movement and an arrangement similar to transverse axial imaging. The essence of the technique was that the observer used a viewing tube which telescopically connected via prisms to the line of sight of the X-ray tube principal axis. In this way, structure at the pivot remained in focus whilst other structure was blurred.

The idea of using a fluoroscopic device combined with a video disc recorder for tomography was developed by Baily *et al* (1974). The addition of the video disc meant that a great number of section images could be reconstructed *a posteriori* simply by reregistering the video data with a variety of shifts applied to the images. For example, if a series of fluoroscopic images obtained with the Grossmann-type linear movement were stored and then these were superposed with zero shift, the usual 'plane through the axis' would be sharply focused. By introducing linear offsets to the data before superpositioning, other planes above and below this fulcrum plane could be refocused. Essentially the equipment achieved seriescopy electronically. However, whereas seriescopy with films was limited to linear tomography, this was not so for electronic seriescopy since circular orbits were almost as easily accommodated by introducing orthogonal offsets to the data. This development was also clearly more convenient than the mechanical methods put forward by Watson and rather eclipsed them especially as the final section images could be reconstructed at will after the patient had gone. Patient dose could also be reduced. Watson's method nevertheless retains that beautiful elegance which often characterizes electromechanical physics. Baily was one of several workers in the 1970s who went on to use fluoroscopic systems for performing not only axial transverse tomography (following the geometry used—with film—by Takahashi) (Baily *et al* 1976) but also digital transaxial CT (Baily 1979).

3.6 The scene changes; from Frank and Takahashi to computed tomography

We have seen how in the period before 1940 all of the important techniques for body-section imaging with reduced blurring became established. This was the golden period of the subject, beginning with a number of disparate workers who invented new machines and clever principles unaware that there were others thinking along the same lines. By the mid-1930s, ideas were fusing; people were being stimulated by the work of others and it was the time of intense rivalry in claims for priority of invention. In view of the importance of these developments we have chosen to present them at some length.

By way of contrast the War years and thereafter were periods of consolidation in which the early ideas were sharpened up, equipment became commercially available and techniques were more widely available and exploited. The inevitable feedback had begun between clinical demands and technological solutions. Transverse axial imaging became more important than the longitudinal techniques. Curved-plane tomography, the logical extension of the earliest attempts by Mayer to reduce blurring, was invented.

A change was in the air, however, signalled by the pioneering work of Gabriel Frank and Shinji Takahashi on opposite sides of the world. They had begun experiments aimed at true single-section imaging wherein the presence of planes above and below that being imaged were of no consequence. They were on the road to what today we call computed tomography (CT). Of course at these times, computing facilities of the type needed to perform CT rapidly were largely unavailable for the purpose and instead these workers used optical methods to 'do the computing'. It is not known whether Frank's apparatus was developed but in the case of Takahashi there are photographs documenting clinical work. The development of CT is strongly linked with the name of Sir Godfrey Hounsfield (see §§4.6.2 and 5.1). He has commented, 'Previous attempts to produce tomographic pictures were theoretically very inefficient in both collecting data (film) and reconstructing the picture (we are here talking about orders of magnitude inefficiency especially in the picture reconstruction). The elimination of this was a major factor which contributed to the success of CT.' (Hounsfield 1989).

With the wisdom of hindsight we know today that modern techniques for CT, which arose as the logical development of analogue tomography, have largely eclipsed interest in non-computed or classical tomography. Whilst it is untrue to say that this has made no further progress, it has been well and truly overtaken in importance and so we reach the point where it is logical to turn attention away from classical imaging and towards CT which is the subject of the second half of this book. It would be fascinating to know whether Takahashi and Frank foresaw what would happen to their ideas if the necessary computations could have been made. Whether or not this is the case, their work should surely be thought of as the groundplate of modern CT. That it arose as a logical development of axial transverse tomography is certain and in this sense we can confidently trace the origins of modern CT imaging methods back to those early days of lone pioneers.

Part 2

Modern History

Chapter 4

Pioneers Towards Computed Tomography

X-ray CT has been hailed as possibly the greatest innovation in radiology since the discovery of X-rays themselves for which Wilhelm Röntgen received the first Nobel Prize for Physics in 1903. In this respect the year 1972 has come to be regarded as the birth of CT since it was in this year that the EMI scanner, the first medically useful X-ray CT scanner, was announced (Ambrose and Hounsfield 1972, 1973, Hounsfield 1973, NS 1972). Hounsfield's genius lay in his ability to overcome the practical difficulties of developing experimental CT and implementing it commercially. The work built on the earlier experiments and theoretical work by A M Cormack and for their contribution to medicine they were jointly awarded the 1979 Nobel Prize for Physiology and Medicine (Hounsfield 1980, Cormack 1980). As might be expected, however, the development of CT has origins which stem back many years to unconnected pieces of research which with hindsight appear prophetic and profound (table 4.1). These are the subject of this chapter. In chapter 5 the story will be opened out with a look at the many ingenious inventions which capitalized on the earliest work.

Table 4.1 Some of the principal landmarks leading towards computed tomography.

Year	Worker; invention
1917	Radon; mathematical solution for reconstruction from projections
1940	Gabriel Frank; pioneering patent (optical single-section tomography)

Table 4.1 *(continued)*

Year	Worker; invention
1945	Shinji Takahashi; experiments with manual and optical methods for back-projection
1956	Cormack; first consideration of X-ray CT in Cape Town
1956	Bracewell; reconstruction of the two-dimensional radio sky from projections
1957	Tetel'Baum and Korenblyum; Russian analogue CT in Kiev
1959	Kuhl and Edwards; first laboratory single-section emission tomography scan (21 August)
1963	Cormack; experimental CT scanner
1960–3	Oldendorf; patent for sensitive point brain scanning
1961	Oldendorf; report of his famous experiment (inspired in 1958)
1963	Kuhl and Edwards; emission CT by analogue methods
1965	Kuhl and Edwards; transmission CT with the Mark 2 imager (14 May)
1967	Hounsfield; first work on X-ray CT (for list of patents by Hounsfield see chapter 5, table 5.1)
1967	Bracewell and Riddle; paper on the convolution and back-projection method of reconstructing from projections
1967	Taylor; reconstruction of radio data from lunar occultation measurements
1968	Kuhl and Edwards; Mark 3 imager
1968	de Rosier and Klug; reconstruction of the ribosome structure from electron microscope projections
1969	Tretiak *et al* ; reconstructions using Fourier space
1970	Crowther; description of the matrix reconstruction method
1970	Vainstein; paper on back-projection reconstruction technique
1970	Swindell; optical tomography
1971	Ramachandran and Lakshminarayanan; paper with the convolution and back-projection method and simulated X-ray CT data
1971	Goitein; simulation of X-ray CT projection data and reconstruction of maps of X-ray linear attenuation coefficient including Cormack's data
1971	Bates and Peters; demonstration of how to reconstruct CT data
1971	First clinical X-ray CT scan with EMI machine (1 October)
1975	Kuhl; Mark 4 imager
(1979	Hounsfield and Cormack; joint award of the Nobel Prize for Physiology and Medicine)

4.1 Oldendorf's experiment

In January 1961 the neurologist W H Oldendorf (born 1915) from the University of California at Los Angeles (UCLA) reported a historic experiment. He was concerned that the very slight variations in the radiodensity of the cranial tissues made it quite impossible to investigate them by non-interventional radiology. The problem is worsened by the containment of these tissues in the radiodense skull. However, Oldendorf, by radiographing a fresh human brain (excised from skull), showed that, even with the skull problem removed, very little useful information could be obtained on brain tissue. It had been known for some 40 years that introducing contrast media could assist visualization of structure (see appendix 1). For example, by introducing air into the ventricles, their structure became visible and, by introducing contrast to the lumen of the blood vessels, further brain structure was perceivable. Both of these practices were traumatic and dangerous and had evolved very little since their introduction. Oldendorf (1961) commented briefly on planigraphy but noted that its clinical utility was again largely confined to the thorax where contrast was generally greater than elsewhere in the body. It was thought to be of little use in the brain.

Oldendorf realized that the problem was one of recording very small changes in signal against large noise backgrounds and this led him to an argument for a proposed solution and a 'home-made' experiment to confirm his belief. He thought that, if a narrow beam of X-rays could be made to interrogate the brain, whilst the brain rotated about a point situated along the beam line, the DC component of the recorded signal would be characteristic of the radiodensity at that centre of rotation. All the other structures of the head would contribute high-frequency signal which could be rejected by a bandpass filter. If the beam were now translated slowly through the rotating head, the point of investigation (the centre of rotation) would sweep slowly through the head. Thus, by recording the low-frequency signal, a trace of the density of structure along this line would be built up. The modulation of the DC component was at a frequency which was a function of the rate of translation of the centre, the diameter of the beam and the abruptness of the change in density along the line being scanned. All other discontinuities in the plane of investigation would modulate the signal at a frequency in excess of twice the rotation rate. Hence it was necessary to have a fairly slow rate of translation compared with the rate of rotation.

Oldendorf wrote, 'At the time of writing, no biological system has been studied by this method. It may indeed prove to be totally useless in such a nearly homogeneous system and is presented here only as a possible approach.' On the contrary his thinking is now much cited as a precursor for CT.

He constructed a novel experiment. He mounted a model train wagon on a piece of H0 gauge railway track which was firmly attached to a gramophone turntable. The railway truck was pulled along the track by a string attached to the hour hand of an alarm clock. The turntable rotated at 16 rev m^{-1} and the truck translated at 80 mm h^{-1}. On the truck he placed a 'brain phantom' comprising two rings of iron nails to simulate the noise-inducing effect of the skull. Towards the centre of the head he placed another single iron nail and a little offset from this, an aluminium nail (figure 4.1). When the phantom was translated without the turntable rotating, a linear scan of the recorded count rate was a noisy trace from which it was impossible to determine the presence of the inner two nails. When, on the other hand, the phantom was translated with the turntable rotating, the signal from the two nails was clearly discernible (figure 4.2). He also conducted similar experiments with a continuous ring of lead to simulate the skull and once again internal structure could be detected.

The importance of Oldendorf's experiment was in showing how, in order to make a measurement of the radiodensity at a point, it is necessary to uncouple the effects of all the other points in the same plane. This was achieved by arranging that the signals were at different frequencies. The trace which was developed by this method of scanning was related to the density of the absorbing phantom traversed by the centre of

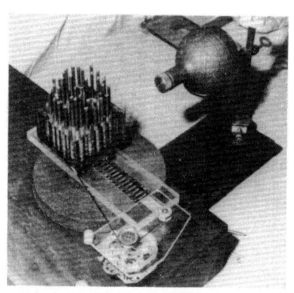

Figure 4.1 A photograph of the arrangement known as Oldendorf's experiment. The source is to the upper right of the picture and the nail phantom is shown on its truck on the railway track. The clock mechanism is to the lower part of the picture. The whole arrangement is mounted on a rotating turntable. (From Oldendorf (1980).)

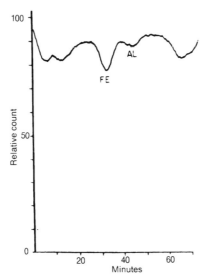

Figure 4.2 The signals from Oldendorf's experiment. The horizontal axis is time and hence proportional to distance moved by the phantom. The vertical axis is the recorded X-ray flux. The signals show the presence of two centrally placed nails inside the ring of externally placed nails. (From Oldendorf (1961). © 1961 IRE (now IEEE).)

rotation. Oldendorf did not, in this paper, take the work further to generate either projections or reconstructed images but the germ of the idea was there. Seen another way, the experiment has analogies with the philosophy of section imaging by classical methods in that the detector was receiving a 'steady' signal from the material at the centre of rotation whilst all the other structures were 'blurred out' in the sense that their signals were averaged out during the measurement.

Oldendorf (1960–3) embodied his ideas in a very thorough patent. In this he clarified the relationship between the linear traversal speed and the angular rotational speed. The former was 'slow' in the sense that at least one revolution occurred during the time a point was moved a distance equal to the diameter of the beam. In a typical embodiment this was 1.6 mm and the revolution was 16 rev min^{-1}. That this condition was easily satisfied is evident from the translation speed set at 80 mm h^{-1}. Oldendorf used a 10 mCi source of ^{131}I directed at a sodium iodide detector coupled to a phototube. By this combination of movements, coupled to the long time constant of the recorder, a unique and discrete region in the opaque object was defined by the intersection of the beam and the axis of rotation. This was isolated and observed whilst all other portions of the object were rejected by virtue of their transverse motion across the line of sight.

In figure 4.3 a typical experimental result from Oldendorf's equipment is shown in some detail in order to make the method really clear. The object 40 is of uniform radio-opacity except for three regions 42, 44 and 46 which are more opaque. The beam 16 was directed at the axis of rotation 20 and impinged on a detector 22. The direction of traverse of the object relative to the source and detector was along the straight line 32. In this particular geometry, the more opaque regions at no time shield each other in the beam and hence, even if the object did not rotate, they would be separable as in the first trace in which the horizontal axis represents position along the linear traverse. The large depressions in the count rate correspond to these three regions coming into the beam. The level 48 corresponds to the unattenuated beam. The depth of the depressions is a function of their radio-opacity and their widths correspond to the linear dimensions of the three regions. However, if we were to imagine that the three regions had been in a straight line parallel to the beam, they could not have been separated in this way. Hence translation without rotation is generally useless. The second trace in the figure is the count rate at the detector with rotation of the object as well as translation. It may be appreciated that every rotation the object passes twice through the beam and causes a large local depression in the count rate. For example the trace at 64 corresponds to when the centre of rotation is not in the object at all. Hence the rate is generally high except when the object sweeps past twice per revolution. 66 corresponds to when the centre of rotation enters the object. Here the general count rate is lower and is further modulated

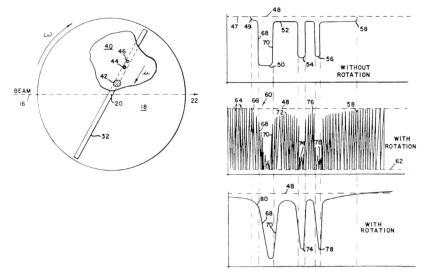

Figure 4.3 A typical experimental result from Oldendorf's experiment. See text for detailed description. (From Oldendorf (1960–3).)

to very low values every time the very dense regions sweep the beam. Around 70 the centre of rotation is in the dense region 42. By 72 it has passed by and the centre of rotation is between regions 42 and 44. The centre of rotation enters 44 and the trace becomes as at 74 and so on. . . .

When the trace in the centre is passed into a recorder with a longish time constant, the trace actually recorded becomes as the third shown. Here the high-frequency components have been smoothed out to leave a simple trace and it is easy to see how the parts of this trace correspond to the passage of the centre of rotation selectively through the regions of the phantom along the line 32. Once again the depth of the curve is proportional to the local X-ray absorption whilst the width is simply the linear dimension of the regions as interrogated by the beam.

In the patent, Oldendorf provided for many different embodiments. Specifically he described equivalent methodology by which, instead of recording the directly transmitted beam, the scattered beam at some fixed angle could be recorded by a similar detector. He also drew equipment whereby the object remained stationary and the source—detector combination rotated. With this arrangement some means of providing the relative translational movement was also required. This could be for example a slow translation of the whole gantry relative to a completely stationary patient.

Interestingly he noted that there was no formal requirement on the rotational movement other than it should be 'fast' in relation to the translation. Specifically it did not need to be continuous rotation at constant angular velocity. Indeed figure 4.4 shows an embodiment in which the rotation was through only a limited angle and was in fact a rocking movement rather than a constant-velocity movement. Provided that the oscillation was of a small period even the two stationary points per period did not matter. Equally the limited angle did not matter.

Finally he drew equipment whereby there was no fast motion but instead the electron beam was directed at a circular target with high angular velocity, the scattered beam being recorded.

Oldendorf explained how two-dimensional images could be generated by sequentially introducing small shifts in the linear scan line. By scanning in a raster pattern a complete two-dimensional image could be constructed. Moreover by moving the plane of the raster scans, three-dimensional imaging was possible. There were no such examples in the patent and it is not clear whether Oldendorf realized these ambitions in practice.

Some historical perspective on Oldendorf's work is provided by the account in his own book (Oldendorf 1980). He has attributed the concept underlying his experiment, namely that high-frequency contaminating signals may be separated from the wanted low-frequency signal, to a chance meeting in 1958 with a young engineer at UCLA who was

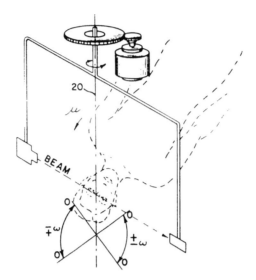

Figure 4.4 A second method of obtaining the radiodensity along a line free from contamination from other points. The source and detector are aligned and rock to and fro through an angle ω. At the same time the patient is slowly translated at speed μ through the beam. The low-frequency signal arises principally from the *point* at which the beam intersects the axis of rotation. (From Oldendorf (1960—3).)

working on a project to develop automated apparatus for the rejection of frost-bitten oranges. These develop dehydrated sections which should be revealed radiographically. Oldendorf hypothesized apparatus (see chapter 6, figure 2 in his book) and from this came the idea for his experiment on radiographing sections of the brain.

It turns out that an earlier patent (Jacobs *et al* 1955—9) 'almost' invented Oldendorf's experiment. This was a patent for a device to examine rocket projectiles and the like for packing flaws when filled with explosives. The projectiles were examined by spinning them about their central axis in a beam of X-rays directed from a tube to a small detector. Variations in the detector output were expected to be due to industrial flaws. The bulk of the patent was concerned with a neat method by which projectiles could be automatically scanned at several layers along their length. However, the axis about which they were spun was not itself translated through the plane of interest and thus the mechanism fell short of what was required to implement Oldendorf's 'sensitive point' method.

Despite his success in obtaining a patent, Oldendorf's approaches to commercial X-ray manufacturers for support were not fruitful. Their lack of interest was not so much a matter of technical difficulty but simply based on risk economics. It is worth requoting a remark in a letter to him from one of the world's major X-ray manufacturers: '. . . even if it could be made to work as you suggest, we cannot imagine a significant market for such an expensive apparatus which would do nothing but make a radiographic cross-section of a head.' This comment led to Oldendorf doing no further work and only returning to CT in 1972. Oldendorf was a clinical neurologist who had the perseverance to develop his ideas in physics with home-made apparatus. He has more recently received many honours for this contribution to neurology and to the physics of CT.

4.2 Russian computed tomography

In §2.3 the pioneering patent from Gabriel Frank was reviewed. This patent issued in 1940 described the essential features of analogue reconstruction of transverse axial sections from projections. The method was based on photographic recording of what today we would call the sinogram and an optical method of producing images. Frank's equipment would have produced only back-projected images since there was no provision for filtering the projections. Later workers noted this and corrected the method (see §2.3). We have also seen how Takahashi was working along similar lines (see §2.4.1).

In the late 1950s, several workers in Kiev published papers which showed that they had solved both the theoretical problem of CT and had developed a method of implementing the result experimentally. The 'discovery' of these papers by workers at the University of Arizona was announced in 1983 (Barrett *et al* 1983). In 1957, Tetel'Baum (1957) developed the equation which is usually referred to as the inverse Radon transform. For less mathematically inclined readers, this is the equation which allows the distribution of X-ray attenuation coefficient to be deduced from its line sums at different orientations about the body. It was first proposed by Radon (1917). A paper on distribution functions by Cramer and Wold (1936) also proved that a distribution is determinable from its projections or moments. In Tetel'Baum's paper the theory is worked out for the case in which the X-rays are in parallel beams although he notes that the fan beam case would admit to a similar solution. He shows a schematic diagram of how the projections might be formed on a continuously moving film and the concept is very similar to that proposed by Frank.

A year later, Korenblyum *et al* (1958) reworked the solution for a fan X-ray beam configuration. This paper, however, also described an experimental arrangement for reconstruction. The image was recorded by rotating the object in a fan beam of X-rays onto a laterally translating film (figure 4.5). This film was then wrapped around a drum and projections from the sinogram were converted into an electrical signal by a television-based analogue system (figure 4.6). Unlike Frank's reconstructor, however, proper provision for implementing the inverse Radon

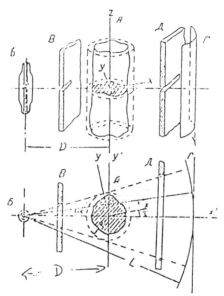

Figure 4.5 The apparatus of Korenblyum *et al* for CT. 6 is the X-ray source, and B is a slit defining the narrow beam and determining the shaded section of the body A. There is also some exit collimation and film for detector. The upper figure is a profile and the lower a plan view. (From Korenblyum *et al* (1958).)

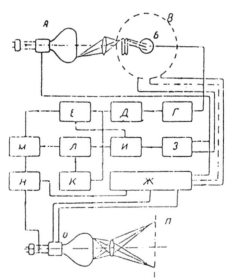

Figure 4.6 The analogue reconstructor for the Russian CT. The X-ray image (called a shifrogram in Russian (sinogram)) is 'read' by an apparatus consisting of a television monitor A and photomultiplier 6. The sinogram is fixed on the rotating drum B. The function of the rest of the figure is described in detail in the Russian paper. (From Korenblyum *et al* (1958).)

transform (that is projection filtering) was incorporated by analogue methods. They stated reconstruction of a 100 by 100 pixel image took 5 min and wrote, 'At the present time at Kiev Polytechnic Institute, we are constructing the first experimental apparatus for getting X-ray images of thin sections by the scheme described in this article.' Tetel'Baum's paper indicated the aim was medical diagnosis and that the method would also find application in non-destructive testing. Tetel'Baum (1956) published an even earlier paper and precise dating of this work is difficult.

These papers were entirely overlooked in the English literature until attention was recently drawn to them. Both have now been translated into English by Professor Barrett and Professor Boag respectively. Hawkins (in Barrett's department) coded Korenblyum's algorithm and found it performed satisfactorily if a little slowly. In 1983 it was reported that no further Russian references to reconstructed images had been found and this still appears to be the case (Barrett 1987).

4.3 Kuhl and Edwards' experiments

Clinical radioisotope scanning was introduced as a diagnostic tool in the late 1940s and became very widely developed in the 1950s. A collimated detector was scanned in a raster fashion to record the photons emitted in its direction from a radiopharmaceutical which had localized in the region of interest. The early detectors were Geiger–Müller tubes, later superseded by scintillation detectors comprising a small sodium iodide crystal and a photomultiplier tube. In the later 1950s the Anger or gamma camera competed with scanning single detectors. For a brief review see Webb (1988).

The scanner generated an image which had similar properties to that of a plane X-radiograph, namely that the recording (in this case count density) at some point was a function (complicated by photon attenuation) of a property (in this case the activity) of all sites at those depths directly in line with the axis of the detector. For high-energy isotopes and thin objects, the count was simply proportional to the summed activity. For lower-energy emitters and thicker bodies, both attenuation and solid-angle effects modified the relationship. During the scan the 'tubes of activity' contributing to the points in the image were parallel to each other and normal to the scan plane. Only in this respect did the scan differ from the essentially fan-beam arrangement of X-radiography. The formal similarities between scanning a radioisotope distribution and taking a planar X-ray prompted Kuhl and Edwards (1963) to ask whether the techniques which were employed in body-section (X-)radiology could be applied to isotope scanners. We shall look at the history of emission tomography in some detail in chapter 6 but the work of Kuhl and

Edwards is reviewed here because it logically groups with that of other workers in this chapter as a precursor to X-ray CT.

Firstly they adapted the oldest of X-radiology's three-dimensional techniques, stereoscopy. They formed two scans with the axis of the detector inclined to the normal to produce a stereo pair of scans. The result was not very successful and they attributed this to the fact that X-ray stereoscopy relies on resolving small changes in the two images. The poorer resolution of isotope scanning seemed to invalidate this approach.

Secondly they adapted the principles of section imaging by which one single plane remains in focus whilst the others are blurred. They made a series of scans of the same object with the detector angled at 30° to the vertical at the four principal compass points. When the resulting scans were superposed with appropriate linear shifts, the structures from a single plane were superposed with other structures not reinforcing and seen as blurs. By varying the shifts, planes at different depths could be focused. They built a calibrated device to achieve this. This is reminiscent of the seriescope devised by Ziedses des Plantes (see § 1.3.2). There was no requirement for the plane of interest to be localized ahead of time. Their results showed that this was more successful than stereoscopy. The strategy for longitudinal section imaging was complete in Kuhl's notebook as early as 16 October 1958 (Kuhl 1989).

Kuhl and Edwards are, however, most cited for their experiments on *transverse-section* emission tomography. They referred to the work of Vallebona and of Kieffer and set out to adapt this idea to isotope imaging. They arranged for the detector(s) to make a line scan *in a single plane* at a number of angular orientations up to 24 in 360°. The way this was done can be seen by reference to figure 4.7 showing the Mark 2 scanner. The images in all their experiments were formed by modulating the brightness of the spot on a cathode-ray tube, the position of the spot being a one-to-one relation to the position of the scanning detector for the experiments described. For transverse-section scanning, the image was also built up on the tube but, instead of a point correspondence, a line sweep was arranged with the direction of travel of the line being normal to that of the detector. The line sweep corresponded to the straight line between the opposing detectors. At a fixed orientation the line translated across the screen and then (as the detectors moved to a new angular position) the line sweep followed suit.

The image they generated was what we today would call a back-projection, the signals being unfiltered. Nevertheless the experiment generated the approximately correct spatial positions of sources in a phantom, even if the images were somewhat blurred. They noted that the images improved as the number of scan lines was increased but even so did not use more than 24. It is interesting that they described their experiment as analogous to those of Vallebona and Kieffer who used

Figure 4.7 A time lapse photograph of the double-headed scanner used by Kuhl and Edwards to perform transverse-section emission (non-computed) tomography. The pair of detectors are shown at two positions in their linear travel for two orientations of the gantry. By sweeping a spot whose intensity was proportional to the recorded count rate and arranging that the sweep followed the positions of the detectors, a simple back-projected tomogram was recorded. (From Kuhl (1968) in *Fundamental Problems in Scanning* ed A Gottschalk and R N Beck. Courtesy of Charles C Thomas, Publisher, Springfield, Illinois.)

X-ray beams which were angled relative to the plane of section. It is the confinement of the scan lines to the single plane that enables us to regard Kuhl and Edwards' experiment as an important precursor of emission CT. Of course the 'computer' involved was an optical integrator rather than a digital computer! In 1963, ^{131}I and ^{198}Au were the radionuclides of choice for imaging deep tumours. Kuhl and Edwards predicted an interest in thyroid and liver tomography. This work established these workers as pioneers in the history of emission tomography. It is amusing to note (Kuhl 1968) that some workers at the time thought there was little interest amongst neurosurgeons (for whom the technique had primary impact) in three-dimensional display and that all that was required was to give them the coordinates of a tumour and their imagination would do the rest in terms of relating this to cerebral structures (such as the midline). Kuhl was one of the first to point out the difference between creating three-dimensional data (for which he thought there was a need) and displaying data in three dimensions (about which he was more sceptical). Even today people will still argue about the best ways to display tomographic data (see also appendix 3).

Several versions of the Kuhl scanner were built. The Mark 3 scanner had four detector assemblies and could therefore perform simultaneous rectilinear scanning in a variety of orientations as well as improved transaxial tomographic imaging (Kuhl *et al* 1974). The Mark 3 imager was initiated in 1965 and completed in August 1968 (Kuhl and Edwards 1969) (figure 4.8). The time required for data capture for the Mark 3 was some 3 min per section. The Mark 4 system (figure 4.9) (built between 1969 and 1975) increased the number of scanning heads to a rotating set of 32 and a transverse-section scan with resolution 16 mm was achievable in 50 s. However, it is perhaps the Mark 2 scanner which was of pioneering importance. The Mark 2 scanner had been fully designed by July 1959, began to be constructed in September 1959 and was in use for emission tomography in the summer of 1963 (figure 4.10). The Mark 2 scanner was used as early as 1965 to make transmission CT images by replacing one of the detectors with a small collimated source of ^{241}Am (Kuhl *et al* 1966). It is interesting to observe that the idea for Kuhl's original *emission* section scanner derived from *transmission* X-ray axial transverse tomography and then contributed to the further development of *transmission* CT in this way (Kuhl *et al* 1974). Of course with only a weak source of gamma rays and the collimation of the emission section scanner, only fairly crude transmission CT sections were generated and they look not unlike those achievable with axial transverse X-ray tomography. They were, however, capable of assisting with the understanding of orientation and geometry when viewed alongside emission section images. Interestingly today this problem of registering images from different modalities is still not regarded as adequately solved. In their paper, Kuhl *et al* (1966) credit Mayneord (1952) with the original idea of outlining organs with

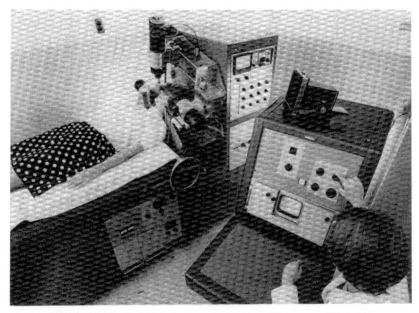

Figure 4.8 The Kuhl Mark 3 scanner in 1968. (Courtesy of D Kuhl.)

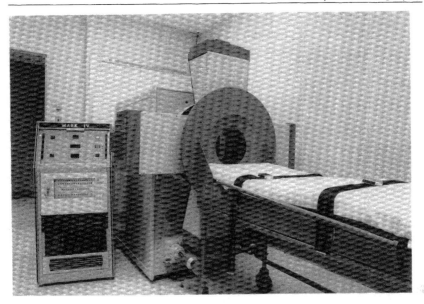

Figure 4.9 The Kuhl Mark 4 scanner in 1975. (Courtesy of D Kuhl.)

Figure 4.10 Dr David Kuhl with the Mark 2 scanner in 1963. (Courtesy of D Kuhl.)

the use of a transmission counter. Kuhl *et al* used the transmission planar scan to assist with understanding emission scans with the equipment in its scanning mode. The transmission data and emission data were simultaneously captured via the use of single-channel analysers to separate the signals.

As the newer emission scanners came into service, the data reconstruction became more sophisticated and simple unweighted back-projection or half-back-projection methods were abandoned in favour of methods which approximately accounted for photon attenuation in tissues (Kuhl and Edwards 1970, Kuhl *et al* 1972). Ell (1982) has remarked that the development of section scanning by Kuhl was an idea which in some ways was ahead of its time in that, at least at the start, the analogue reconstruction techniques did not do justice to the experimental data.

Dr David Kuhl has recently kindly supplied a very detailed report on his contribution to CT (Kuhl 1989) from which it has been possible to pinpoint more clearly the exact dates for fundamental landmarks. This report also confirms just how far reaching has been Kuhl's contribution. His development of photorecording for radionuclide imaging as far back as 1952 should not be overlooked. At first this replaced the paper tapper for rectilinear scanners but it became a vital part of the analogue display for section imagers. He wrote, 'This method of photorecording, together with the introduction of the large crystal scintillation detector, the focused collimator, and pulse height analysis comprised the essential developments that made possible the early radionuclide scanner as a useful clinical instrument.' Kuhl's motivation for emission tomography stemmed from wanting to replace autoradiography as a means of studying internal organ physiology with radioactive indicators. His interest started in high school in 1946 and continued in his spare time through 1952 to 1956 whilst in medical school at the University of Pennsylvania, spare hours being devoted to the construction of a rectilinear scanner using the photo-recording method.

After National Service between 1956 and 1958, Kuhl began working out the concepts of transverse-section scanning, spurred on by the hope of being able to quantitate cerebral blood flow with section images. He joined forces with Mr Roy Edwards who was in charge of the engineering shop in the Department of Radiology. Kuhl states, 'My notebook shows that I had developed the transverse-section scanning method fairly well in theory by 25 January 1959, at which time I had decided to perform transverse-section scanning at interval angles of 20° using the superimposition of a modulated line of sight photographed from the screen of an oscilloscope. I had already shown to my satisfaction that a fairly accurate cross-section picture could be reconstructed by back-projection and superposition of data obtained by scanning in paths tangent to the circumference of the cross-section. My first "scanner" was a turntable for the radioactive object positioned directly in front of a Bridgeport Milling Machine to which was attached the traversing scintillation detector. Roy and I produced the first successful emission transverse-section scan with this apparatus by working non-stop through the night of 21 August 1959. . . . On 28 November 1959 I made the first public presentation of my work . . . in a lecture presented at the US Naval Medical Centre, Bethesda, Maryland. . . . By mid-1959, I had designed the basic configuration of a scanning machine which was to become the Mark 2 scanner. . . . We built the machine in the basement of the University Hospital and did not finish it until 1963. This is the instrument with which we performed the first clinical emission computed tomography and the first transmission transverse tomography of man, as well.'

Kuhl affirmed the independence of his iterative reconstruction methods, including the 'orthogonal tangent correction', developed between 1966 and 1972 for use with the Mark 3 and Mark 4 scanners from the work of others in this field. In particular he wrote, 'We now know that, at approximately the same time, similar insights concerning reconstruction algorithms were being developed independently for other purposes in laboratories other than those of either Kuhl or Hounsfield. . . . We developed this algebraic iterative algorithm without knowledge of Dr Hounsfield's work which was in progress at approximately the same time.' (see also §4.4).

As early as 13 June 1955, Kuhl had produced transmission rectilinear scans using a ^{131}I source coupled to a scintillation detector, spurred on by Mayneord's work. Regarding the use of the Mark 2 scanner for making transmission CT scans Kuhl dated the first-ever scan to 14 May 1965: 'This was the first transmission CT picture of a patient. . . . To the best of my knowledge, this was the first such image ever made, clearly prior to Hounsfield's work. . . . The procedure was very similar to that later employed in Hounsfield's EMI scanner, except that I used a ^{241}Am transmission photon source rather than an X-ray generator and, of course, our reconstruction strategies were more primitive then. I understand Dr Hounsfield's first experiments with mock-ups in 1967 also used a ^{241}Am source and a scintillation detector with this same scanning mode. . . I reasoned then that such transmission CT pictures could be useful in radiotherapy planning. . . . In retrospect, it is very likely that a modification of our 1965 transmission CT scanner by only adding a more intense radiation source and using a more effective reconstruction algorithm would have resulted in an imaging system capable of the same kind of attenuation pictures produced by the first commercial transmission CT scanners. However, my principal interest, then and now, was quantitative emission CT. . . . I discontinued my efforts for further development of transmission CT instruments after the announcement of the much more advanced commercial unit by EMI in 1972.'

There seems no doubt that Kuhl happily acknowledges that the EMI machine was considerably more advanced than what he was able to achieve at that time. However, he recorded, 'I have little knowledge as to how any of the published work on reconstruction tomography by Kuhl and coworkers prior to 1973 might have been useful to Dr Hounsfield in his research and development activities. I have never had a conversation with him, nor have we ever exchanged correspondence directly. In his first most comprehensive published paper on the EMI scanner Dr Hounsfield references none of our prior papers. This was surprising since the editorial immediately preceding this paper stated that the mechanical system of Dr Hounsfield resembled that of Dr Kuhl. This omission was disappointing to me since a year before I had received a letter from the EMI Central Research Laboratories stating, "As I mentioned to you, we at EMI have been following very closely your work and there is obviously some similarity."' The similarity between these pieces of work was noted in press many times and Kuhl's (1989) document lists examples. Perhaps bearing in mind the historical precedent accomplishing the announcements of classical tomography the situation is not really surprising. What should emerge in our minds is that both these workers (and many others) made outstanding personal contributions of far-reaching consequence.

4.4 Pioneers in mainly non-medical computed tomography

During the 1960s a number of people were working on the CT problem for reasons unconnected with medical imaging. With hindsight their work has been much cited as pertinent to medical X-ray CT. Two fields of research, in which essentially the identical tomographic problem arose, were radioastronomy and electron microscopy. Brooks and Di Chiro (1976) have written, 'One striking fact is the small amount of cross-fertilization between the various disciplines. This is perhaps inevitable in view of the extent and specialization of scientific literature.'

As early as 1956, Bracewell (1956) had shown that two-dimensional maps of structures in the radio sky were derivable from one-dimensional linear measurements (or projections) obtained with a radiotelescope. Subsequently Bracewell and Riddle (1967) produced a mathematical algorithm for solving this problem which did not require the computation of Fourier transforms. (At that time of course taking transforms required much more computer time than at present; so this development was regarded as a distinct advantage.) The equations, in what has become a classic citation, are those of the now familiar convolution and back-projection real-space method in which the ramp filter plays its vital role in yielding tomograms in which the $1/r$ blurring function (see §2.3) has been effectively removed. (Readers unfamiliar with this problem may wish to consult appendix 4 at this stage.) They used this method to derive a two-dimensional map of lunar radio brightness from measurements with the 9.1 cm interferometer at Stanford. For this problem, hitherto unsolved, 37 line scans or projections were utilized. That this was adequate was shown from a formula, arising during the derivation, giving the minimum number of views which must be measured in terms of the diameter of, and the highest spatial frequency in, the object to be reconstructed.

Reconstruction from projections has been fruitfully applied to a number of problems in astronomy. Instruments for measuring the radiobrightness of the sky sometimes lack the required spatial resolution. If the moon moves across the region of interest in the sky, the radio data from such lunar occultations act as projections or profiles and may be used to reconstruct the two-dimensional distribution of the brightness of a radio source. The method makes use of the fact that the direction of the path of the moon across the sky varies from day to day. Bracewell and Riddle (1967) submitted their paper to the *Astrophysical Journal* on 14 April 1967. 3 days later a paper by Taylor (1967) was submitted to the same journal (and subsequently published alongside the other paper in the November issue), describing the use of lunar occultation data for reconstruction. Taylor took measurements with the 140 ft telescope at the National Radio Astronomy Observatory at Green Bank, West Virginia. Six one-dimensional profiles of the radio source 3C 192 derived from occultation observations at a frequency of 256 MHz were used to reconstruct the two-dimensional map of the galaxy. With the largest diameter of the galaxy being some 220′ and the largest angle between adjacent profiles being 32° (0.56 rad), a resolution of 60′ resulted in the two-dimensional map. This resolution was calculated with the by then well known result that a finite maximum observable spatial frequency led to a non-zero minimum spatial resolution. He used the direct Fourier method based on the central-section theorem that the one-dimensional Fourier transforms of the projections were central diameters in the Fourier space version of the two-dimensional galaxy. The data were arranged on a 25 by 25 pixel array; two-dimensional Fourier transforms took 8 s on a CDC 6400 computer.

In Cambridge, a group of electron microscopists developed similar and complementary theory to solve a different problem. They were interested in elucidating the three-dimensional structure of bacteriophage and viruses from their two-dimensional electron micrographs. These had to be digitized with a computer-controlled film scanner. During this process both the amplitude and the phase information was preserved unlike during X-ray diffraction measurements in which phase is lost and this allowed reconstruction in three dimensions. To obtain views of the structure from different orientations, the table in the electron microscope was tilted to a number of different positions. Because the depth of focus for each measurement was several thousand Ångströms the data were essentially two-dimensional projections of the three-dimensional structure.

The crystallographers were helped by the inherent symmetry in many of the structures in which they were interested which enabled a fewer number of views to be used than would have been the case for structures with no symmetry. For example the T4 phage tail of interest with helical symmetry required only a single view whilst a ribosome with no symmetry required of the order of 30 views. Minimizing the number of views was important because of the potential destruction of the sample in the electron microscope.

At first the three-dimensional structures were reconstructed by Fourier methods (de Rosier and Klug 1968) but other techniques were developed. Crowther *et al* (1970) worked on matrix inversion methods which produced results in good agreement with the Fourier route. Gilbert (1972b) formally derived the mathematics of back-projection including the constant (also given by the Russian worker Vainstein (1970)) which had to be subtracted from the back-projected image in order that the picture so formed was quantitatively correct. In addition to this, Gilbert showed the limitations of back-projection by deriving the relationship between the back-projected image and the result from filtered back-projection (in which the $1/r$ function has been deconvolved). Gilbert then went on to derive the real-space convolution and back-projection method and used it to work out the structure of tobacco mosaic virus protein. He noted most clearly that his reconstruction formula was exactly the same as the result derived differently by Bracewell and Riddle (1967) and by Ramachandran and Lakshminarayanan (1971). Bender *et al* (1970) in Buffalo used the algebraic reconstruction technique to elucidate the structure of the ribosome with some 20 Å resolution.

The Fourier space reconstruction technique had also been reported by Tretiak *et al* (1969). They proposed taking a series of X-ray photographs of a rotating subject and using these to reconstruct internal structure. Their conference abstract included the statement of the now well known central-section theorem that the two-dimensional Fourier transforms of the X-ray photographs yielded the sections through the three-dimensional transform of the required object structure. Were American minds concentrating? 2 days prior to this paper, the *Apollo 11* lunar module Eagle had made the first moon landing in the Sea of Tranquillity.

Despite the fact that Radon had essentially given the method of reconstructing from projections as early as 1917, many people took up the challenge. Berry and Gibbs (1970) in Bristol wrote, 'The first question considered is whether a series of projections (such as X-ray shadow-graphs) of a three-dimensional object can be used to infer unambiguously the structure of the object. It is shown that this is the case.' They also proved the central-section theorem by a Fourier method, reduced the three-dimensional problem to the now-familiar stacked two-dimensional problem and derived the Radon inversion formula, describing the result as 'straightforward . . . (but) . . . analytically complicated'. No experimental data were taken but clearly the concept of performing CT was firmly established in a practical way.

In New Zealand, Bates and Peters (1971) independently developed the mathematics of CT. They did this in a neat way which provided the connection between 'classical' axial transverse tomography and true CT. Their final equation showed that the true cross-section was the rho-filtered layergram. If the X-rays in axial transverse tomography were at grazing incidence, the resulting layergram could be Fourier transformed in rectangular coordinates. After multiplication of the result in frequency space by a linear ramp in frequency (the so-called rho filter), the CT slice resulted from inverse Fourier transformation. They showed how such a process could conveniently be achieved optically using two thin lenses and an optical mask. Similar work was performed by Swindell (1970). Around this time the computation of accurate fast Fourier transforms was only just becoming known, explaining why Bates and Peters were concerned with performing some of their processing with a digital computer and some optically. Much of the language of this paper is somewhat prophetic. They wrote, 'We are currently studying modifications of existing transverse tomographic equipment which could produce true layergrams. We also point out that the X-rays could be detected by a scintillation counter with its output fed directly to computer storage. . . . It would be straightforward to program the computer on the basis of the theoretical results presented below . . . to calculate the accurate cross-section.' In California, Rowley (1969) was using this kind of mathematics to solve problems in holographic interferometry.

Notice the timing of these papers. They all pre-dated the public announcement in 1972 of the first medical CT scanner (see §4.6) and much of the work was largely contemporary with this medical development. Image reconstruction from projections has remained of interest to the non-medical community. For example an X-ray telescope on board a rocket going beyond the atmosphere may also record projections of the X-ray brightness across the sky. This has been used to map the Vela supernova remnant for example (Herman 1980). Maps of the electron density in the solar corona have also been reconstructed from projection data from instruments measuring the total electron density along the line of sight over a period of half a solar rotation (14 days) (Altschuler 1979). Projection data have also been used to map the spatial concentration of air pollutants and for problems as diverse as estimating the size of icebergs from shadow data recorded by a moving ship (Herman 1980).

187

4.5 Goitein's reconstruction

Another paper to appear almost prophetically in 1972 and before the announcement of the first medical CT scanner was that by Goitein (1972). This included a consideration of the reconstruction problem for a number of applications, positron tomography, charged-particle tomography, electron microscopy and X-ray CT. Figure 4.11 shows how Goitein expected to reconstruct the distribution of X-ray linear attenuation coefficient from a series of transmission projections. He proposed that a matrix inversion technique be used to give a coarse-scale two-dimensional image which was then refined by a relaxed iterative reconstruction method to give an image with smaller pixels. This philosophy derived from acknowledging the computational difficulty of a fine-scale matrix method. Goitein was given three kinds of experimental data with which to demonstrate his methodology.

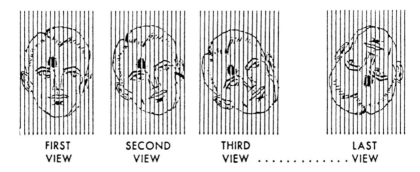

FIRST VIEW SECOND VIEW THIRD VIEW . . . . LAST VIEW

Figure 4.11 One of the earliest published figures showing how a series of measurements comprising projections along parallel paths at different orientations form the basis of reconstruction X-ray CT. In practice of course the rotation would be in a transaxial plane—the 'face' here is used to illustrate the concept. (From Goitein (1972).)

He first became interested in image reconstruction from projections after a stimulating visit to the Donner Laboratory (UCLA) and discussions with Tobias (Goitein 1989). Tobias' interest was in radiotherapy with charged particles and he suggested planning particle therapy would be assisted by imaging cross-sections with charged particles. Goitein took up the challenge with data from J Lyman of a chest phantom recorded with 840 MeV alpha rays at the 184 in Lawrence Berkeley cyclotron. These were coarse data; only 19 viewing angles with 41 elements per view. Figure 4.12 shows the result. The second data set were X-ray CT data of a spherically symmetric annulus, from D Chesler, whom Goitein had met on a visit to the Massachusetts General Hospital (MGH) in Boston. The

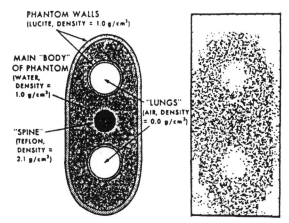

Figure 4.12 (*a*) A chest phantom and (*b*) its reconstruction from 840 MeV alpha-particle projections. (From Goitein (1972).)

single projection was replicated to simulate the views from other orientations. The third data set was donated by A M Cormack, being the same as that used for his reconstruction (Cormack 1964). There were 25 projections each with 19 elements. The result is reproduced in figure 4.13.

Goitein did not construct a scanner himself nor take data but this work was important, being among the first efforts to make mathematically

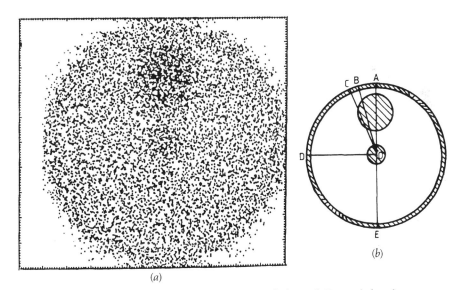

Figure 4.13 (*a*) The reconstruction of some original data of Cormack by the new relaxed iterative technique published by Goitein. (From Goitein (1972).) (*b*) The phantom constructed by Cormack. The shaded areas are aluminium; the unshaded interior area is Lucite. (From Cormack (1964).)

derived cross-sectional images from real data with a well thought-out treatment of the processes involved and their limitations. It was of course done with good knowledge of the crystallographic analogue and the early work of Cormack (see §4.6.1).

Goitein (1989) feels that to a large extent the solution of practical CT was simply a matter of time, being worked on by a number of people simultaneously. The award of the Nobel Prize, whilst justly acknowledging those who finally solved the practical difficulties, inevitably diminished the collective achievements of these other workers.

4.6 Towards the Nobel Prize

The 1979 Nobel Prize for Physiology and Medicine was jointly awarded to Allan M Cormack and Godfrey N Hounsfield for their work leading to the invention of a medical transaxial CT scanner. The reader can do little better than to consult their Nobel Foundation lectures for their own accounts (Cormack 1980, Hounsfield 1980). A brief review now follows.

4.6.1 Cormack's account

Cormack first considered CT in 1956 when he was a lecturer in physics at the University of Cape Town. Part of his time was spent at the Groot Schuur Hospital where he observed radiotherapy planning techniques. He concluded that these could be improved if there were a map of the X-ray linear attenuation coefficient in the slice being planned—what today we would call tissue inhomogeneity corrections. Surprised to find that this problem had not been thought of before, he formulated the mathematics of reconstruction from projections for simple attenuating objects with certain symmetry. He applied this reconstruction technique in 1957 to experimental data with some success. It was not until 1970 that he became aware that the solution had been found in 1917 by Radon. In 1963, together with an undergraduate student, David Hennage, he constructed an experimental scanner and used it to reconstruct asymmetrical sections of phantoms. The phantoms which they used had materials chosen such that their attenuation coefficient ratios were similar to the ratios of uptake of radionuclides between tumour and normal tissue. There was much interest at that time in positron tomography (see §6.4) (Wrenn *et al* 1951). Cormack published the results (Cormack 1963, 1964) but noted that the only real interest in them was expressed by the Swiss Avalanche Research Centre, hoping the method would be of use in predicting depth of snow.

The interest in CT was shelved in favour of other work until the early 1970s. At that time, Cormack became aware not only of Radon's work but also of the work of Bracewell (1956) in radioastronomy, de Rosier and Klug (1968) in electron microscopy and other earlier mathematical work. Cormack also foresaw the possibility of proton CT utilizing the sensitivity of the proton range to tissue inhomogeneity and noting that the CT sections could be used for planning proton radiotherapy. He dismissed contrary arguments of high cost and with Koehler produced proton CT data (Cormack and Koehler 1976), work continued by Hanson and colleagues. It is sometimes remarked that Cormack's main contribution lay in the mathematics of CT but his was also an innovative experimental contribution to winning the joint prize.

4.6.2 Hounsfield's account

Hounsfield was less specific about when he first became interested in the possibility of creating CT data. In his Nobel lecture he simply referred to 'early tests' of laboratory CT. His equipment (figure 4.14) used a gamma-ray source collimated to face a single detector. Between the two a phantom was rotated in 1° steps on a turntable. The scanning movements of the source and detector were provided by a lathe bed. 28 000 measurements were recorded per slice on paper tape. The data took 9 days to collect per slice and 2.5 h to reconstruct on a mainframe computer. A further 2 h was required to display an image from the

Figure 4.14 A much-published figure showing Hounsfield alongside his laboratory prototype equipment for X-ray transmission CT. (From Oldendorf (1980).)

digitized result. By replacing the gamma-ray source by an X-ray source the data capture time was reduced to 9 h. Still, however, only one image per day was possible and in this time some specimens under examination were suspected of changing their properties in view of producing gas bubbles. The machine which eventually went into clinical trial at the Atkinson Morley Hospital took high-resolution images in just 18 s and was first used in 1971. Hounsfield recognized early the applicability of CT data for aiding radiotherapy planning as well as for radiodiagnosis. The design of CT scanners was very soon optimized in terms of their photon utility and, in his lecture, Hounsfield pointed to the next logical development of essentially real-time CT scanning which is still only a reality in a few centres. The company in which Hounsfield was employed (EMI) pioneered the commercial development of both transmission X-ray CT and later nuclear magnetic resonance body-section imaging.

The precise order of events leading to the development of the first clinical scanner are necessarily shrouded in a certain obscurity; what is clear is that the development was a three-sided one with involvement and interest of the EMI Company, the British Department of Health and Social Security (DHSS) and the clinical staff at the Atkinson Morley Hospital in London (Ambrose 1988, Bull 1981). Dr James Bull at the Institute of Neurology and Dr Evan Lennon suggested the meeting between Hounsfield and Ambrose. They met in 1967 and Ambrose recognized the potential importance of Hounsfield's suggestions to the field of neurology. Ambrose was working on ultrasonic imaging of the brain by sectional techniques and it became clear to them both that X-ray CT could probably improve on this. Ambrose provided bottled cut brain specimens for laboratory experiments in Hayes and was delighted to find that tumour, white matter and haemorrhage were all visible on the reconstructed scans. Further experimentation occurred with Kreel on pigs' cadavers and with Ambrose on ox brains and the results were sufficiently encouraging to persuade G Higson, N Slark, J Williams and D Gregory at the DHSS to fund partly the development of a prototype head scanner. The work took about a year, the machine being hand made in the workshop at Hayes where Ambrose and colleagues paid many visits. The first clinical scan was made on 1 October 1971. This was of a 41 year old female with a suspected left frontal lobe tumour. These clinical images have been reprinted on many occasions (see, for example, Ambrose 1977). The project was carefully kept secret until the 1972 British Institute of Radiology Conference in April. It was a well guarded secret because of necessity a large number of people at the Atkinson Morley Hospital needed to know what was going on. When the news broke the scientific community were amazed at what had been achieved.

As well as sharing the Nobel Prize with Cormack, Hounsfield was also awarded the MacRobert Award (often described as the Nobel Prize for

Engineering) in 1972. The evaluation committee wrote, 'There is another aspect of the EMI scanner which is remarkable. In these modern days it is rarely that one finds great developments which are the work of one man. The EMI scanner is different . . . (it) was as much a one-man invention as anything can be these days.' (EMI 1972).

Whether one should connect the invention of X-ray CT with the early developments of classical blurring tomography is perhaps a matter for personal philosophy. The edges and connections of one piece of research with another are always blurred but connections, however tenuous, I believe there are. Very similar thoughts were aired by Ambrose (1977) who wrote, 'One is tempted to wonder what might have happened if the originators of tomography, Bocage, Ziedses des Plantes and Vallebona, had been aware of Radon's treatise and had realized that they had a common problem. One might well be excused for thinking that nothing could have come of it since modern computers, essential for handling the vast amount of information, were so far away in the future. But is not necessity the mother of invention?' The development of the first commercial CT scanners was clearly a case of 'standing on the shoulders of giants'.

Ambrose has noted that EMI and Hounsfield are somewhat reticent about the origins of their work but there must be many who relish the picture painted of Hounsfield walking in the back lanes of England pondering these problems. If connections there be, perhaps it is most appropriate that amongst the honours afforded to Hounsfield was (jointly with Oldendorf) the Ziedses des Plantes Medal.

Chapter 5

Patents for Computed Tomography

In earlier chapters the story of the evolution of body-section imaging has been told. In large part this account has concentrated on developments in analogue qualitative blurred so-called classical tomography. However, we have also seen how a number of pioneering workers foresaw how unblurred quantitative single-plane tomographic imaging could be obtained even though in general they had at their disposal only limited means of realizing their ideas in practice. In particular, before the days of the widespread availability of digital computing techniques, those reconstructed cross-sectional images that were made were necessarily produced by analogue methods and were largely not quantitative. There was limited scope to cope with data deficiencies and in particular to implement correctly the filtered back-projection process which leads to unblurred imaging and properly inverts the process of projection data capture. We recall for example that Frank (1940) patented a CT-like procedure but that the images which equipment based on these principles would produce would have been unfiltered back-projections. Takahashi had also clearly shown how cross-sectional images related to their projection sinograms. It is also fair to say that the methods of capturing projection data were in themselves rather cumbersome, generally being based on the use of film as detector such as in the pioneering experiments described by Takahashi (1957).

With the possible exception of the work in Russia on CT in the late 1950s—and documented evidence of the practical implementation of even these ideas in a medical context is hard to find—it was not until the late 1960s that what today we should recognize as modern CT came to be achieved. This was characterized by a departure from the use of film as detector, by complex mechanical and electrical engineering to implement the projection data capture and by digital numerical computer reconstruction which led to quantitative, spatially accurate sensitive mapping of the X-ray attenuation properties of body cross-sections.

Of all the developments in radiology—and dare one say medical physics in general—with which the general public come to be familiar, the development of rapid-scan CT must surely be one of the most well known. The impact of this development on medical science has been colossal; some would say equivalent to the impact of the discovery of X-rays by Röntgen in 1895 (Burrows 1986). Images from CT scanners often illustrate the colour supplements of national newspapers. Equally the basic principles of CT scanning are generally taught to students of medical physics and to radiologists and radiographers. Hence there is little need to repeat these principles in depth here. However, in a sense, the more popular a scientific development, the less connected to its origin it often becomes, and, whilst it is not too difficult to describe the basic principles of CT, most of us might find it less easy to assign specific developments to their authors with chronology. Also, in simplified reviews of CT, the features of the many developments are sometimes fused into a common story, once more with details becoming obscured. In this chapter we return to the original patents for CT and pull out the key developments. It is hoped that this approach may be seen as complementary to the many already adequate detailed descriptions in the literature of the physics of CT. Important patents are summarized in table 5.1. The CT head scanner was almost instantly accepted as a revolution in radiology, removing the requirement for dangerous encephalography. The usefulness of CT body scanners, however, took a longer while to become accepted (Stocking and Morrison 1978).

Additionally there is something of a difference in approach taken in this chapter. Working out the story of early body-section imaging required some detective skills. Gradually the pieces of the jigsaw fitted together and one can with hindsight link developments in a common theme. With only a few exceptions the workers at the time (at least pre-1935) were largely unaware of what the others were doing. When one comes to look at the modern history of CT, quite the reverse is the case. The impact of the first developments by Hounsfield and Cormack was so great and their work became public knowledge so quickly that subsequent work and patents were documented with due reference to what had preceded them. The purpose here is to sift from the vast literature some key landmarks. To a large extent this will also concentrate on technological and experimental innovations rather than on reconstruction theory which has its own rich literature (see, for example, Herman 1980).

5.1 The Hounsfield–EMI patents

The first apparatus for CT was patented by Hounsfield (1968–72) and EMI Limited with the application filed in August 1968. After explaining

Table 5.1 Some of the modern patents for computed tomography between 1968 and 1980 (in chronological order of application). Many of these patents were assigned to commercial companies.

Date Registered	Granted	Author	Country	Number
23 August 1968	2 August 1972	Hounsfield	UK	1283915
27 December 1971	11 December 1973	Hounsfield	USA	3778614
9 April 1973	11 February 1975	Hounsfield	USA	3866047
5 May 1973	30 March 1977	Hounsfield	UK	1468810
10 May 1973	29 April 1975	Hounsfield	USA	3881110
7 May 1974	16 March 1976	Hounsfield	USA	3944833
9 May 1974	11 November 1975	Hounsfield	USA	3919552
9 May 1974	2 December 1975	Hounsfield	USA	3924131
30 May 1974	24 February 1976	Hounsfield	USA	3940625
20 June 1974	12 July 1977	Hounsfield	USA	4035647
17 July 1974	20 January 1976	Hounsfield	USA	3934142
29 November 1974	28 September 1976	Boyd	USA	3983398
12 December 1974	21 December 1976	Hounsfield	USA	3999073
23 January 1975	31 January 1978	LeMay	USA	4071760
15 May 1975	7 December 1976	Froggatt and Percival	USA	3996467

18 May 1975	28 September 1976	Cox and Snyder	USA	3983399
8 August 1975	10 August 1976	Distler	USA	3974388
27 August 1975	11 January 1977	Hounsfield	USA	4002911
5 September 1975	8 November 1977	Wagner	USA	4057725
9 October 1975	28 June 1977	Mayo and Best	USA	4032761
12 November 1975	16 May 1979	Siemens	UK	1546158
28 November 1975	4 October 1977	Brunnett	USA	4052620
19 March 1976	21 June 1977	LeMay	USA	4031395
7 June 1976	13 September 1977	Taylor	USA	4048503
3 November 1976	21 February 1978	Boyd	USA	4075491
13 December 1976	13 March 1979	Wagner	USA	4144570
30 December 1976	28 March 1978	Froggatt	USA	4081681
28 February 1977	18 July 1978	Lill	USA	4101768
8 March 1977	13 November 1979	Liebetruth	USA	4174481
21 March 1977	17 April 1979	Franke	USA	4150293
25 March 1977	3 October 1978	Froggatt	USA	4118631
11 April 1977	6 March 1979	Richey *et al*	USA	4143273
31 January 1978	27 May 1980	Inouye and Mizutani	USA	4205375
22 August 1978		Boyd	Australia	AUS504935B
18 December 1978	23 December 1980	Lux	USA	4241404
19 May 1980	25 November 1981	Swift	UK	2076250A
4 June 1980	13 April 1982	Kalendar, Linke and Pfeiler	USA	4324978

how a two-dimensional X-ray image collapses three-dimensional information into a form which is too confused to interpret, the patent draws attention to that of Oldendorf (1960–3). This we recall (see §4.1) was a method of sensitive point scanning whereby the signals resulting from the X-ray attenuation at a specific point were separated from the confusing signals from all surrounding structure. Of necessity to generate a whole image in this way would be a very time-consuming task and be associated with excessive radiation dose. According to this patent, Cormack's method was also thought to suffer from an unacceptably large degree of computation and did not allow for problems created by noise.

New apparatus was proposed which could rapidly gather the data needed to reconstruct a map of X-ray attenuation with sensitivity enough to visualize the contrast between soft tissues. The basis for the method was the now well known physics that, if a large number of measurements are made of the attenuation in narrow tubes passing through the body at different angles, the desired image may be reconstructed computationally from these data. In the original apparatus the tube was of square cross-section of side 3 mm. The X-ray detector was a collimated scintillator with counter. For illustrative purposes the rotate–translate method of data capture was described, the absorption in some 100 parallel photon paths being measured at some 400 angular orientations in $0-\pi$. The reconstruction method described was a relaxed iteration whereby the entries for attenuation in a square mesh were adjusted via use of those measurements passing through each element or pixel. Considerable attention was paid to the need to randomize the selection of the tubes from the projection data which passed through each point when iterating for the reconstructed value at that point. This was necessary because of the somewhat correlated nature of measurements in adjacent tubes which, if accessed sequentially, led to poorer convergence. Allowance was also made for the fractional intercept of mesh squares and ray tubes via weighting factors approximated to 20 values which could be looked up during computer calculation.

The first actual scanning sequence described in the patent is one to which one sees little reference elsewhere in the literature (figure 5.1). The X-ray source 6 and scintillation detector 7 face each other and rotate about a point within the body. Measurements are taken at discrete steps about the circumference of the scanning circle 14. In order to produce uncorrelated projection data for all points in the body after a complete rotation, the centre of rotation must move to a new place and another rotation be performed. The rationale is that the spatial sampling is much coarser at the periphery than at the centre and hence the method has higher central definition than peripheral definition. The movement of the isocentre could, for example, be around the circle 15 but arbitrary movements were patented. For example the movement could be a spiral

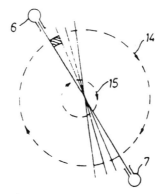

Figure 5.1 The very first CT scanning technique from the very first Hounsfield patent. See text for explanation. (From Hounsfield (1968–72). Courtesy of the Comptroller of Her Majesty's Stationery Office.)

starting at the edge of the outer circle, progressing rapidly towards the centre and then performing a slow spiral in the region of the centre. Today we can see that this scanning method is potentially lengthy and associated with high dose. It is a little reminiscent of Oldendorf's proposal but with the added potential for reconstructing an image.

One practical system proposed is shown in figure 5.2(a). An X-ray tube 20 produces a fan spanning the region defined by 26. The photons are collimated into two channels via collimators 21–24 and impinge on two scintillators 28 and 29. These are coupled via a light pipe 30 to a single

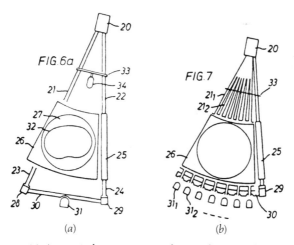

Figure 5.2 (a) A practical CT scanning technique from the first Hounsfield patent. (b) A modification of the apparatus in (a) for more rapid CT. See text for details. (From Hounsfield (1968–72). Courtesy of the Comptroller of Her Majesty's Stationery Office.)

199

photomultiplier tube 31. This is adequate in that the X-rays are arranged to alternate between channels to scintillators 28 and 29 via a chopper 33 which sequentially directs the beam first into one collimator and then into the other and so on.

The patient 32 is arranged inside the aperture 27. The scanning movements comprise a rapid oscillation of the system apart from the X-ray tube such that the channel into 28 sweeps across the patient. Meanwhile the whole apparatus slowly revolves a single turn about the patient. Detector 29 is used to calibrate the measurements. A water bag—a feature of early EMI scanners—was packed around the patient into the space 27, thus reducing the dynamic range required of the photomultiplier—scintillator combination. The attenuator 25 was constructed of material such as Perspex so that the measurement at 29 fell within the same dynamic range. This channel was used to compensate approximately for the spectral drift of the X-ray tube. Thus the equipment contained all the essential features required to collect the projection data for reconstruction. The patent also showed improvements to this design, the first removing the light pipe arrangement and substituting a second photomultiplier for the detector 29. A more ambitious development was the substitution of a number (seven in the figure) of scintillator—photomultiplier detectors spanning the body. This reduces the angular oscillation to no more than one seventh of that of the basic system, speeding scanning (figure 5.2(*b*)).

Regarding reconstruction, in the 1968 patent, provision was made to store the projection data on a tape recorder, to transfer it off-line to a computer and to take the digitized reconstruction further via a second magnetic tape to the display device where image data could be viewed using windows of variable level and width. The latter facility is now commonplace in all digital image viewing facilities but at the time was quite a novel feature. For an account of the difficulties associated with the viewing of analogue reconstructed data see, for example, Kuhl and Edwards (1963). Appendix 3 discusses image display.

Subsequently Hounsfield (1973–7) and EMI patented a modified scanning technique shown in figure 5.3. This became known as the 'rotate-only' method of CT. A bank of *n* radiation-detecting crystals 5 were arranged opposite an X-ray source 1 which was collimated to a wide fan by 4. Thus at any one orientation a large number—*n* was typically 300—of measurements of integral attenuation were obtained. By rotating the device relative to the stationary body 2, sufficient measurements were recorded to reconstruct a map 23 of attenuation within a slice. Interestingly by this date the iterative method of reconstruction had been abandoned in favour of other processing techniques (such as convolution and back-projection). Mayo and Best (1975–7), also from EMI, subsequently patented a mathematical reconstruction technique aimed at avoiding time-consuming multiplications by the use of Walsh functions.

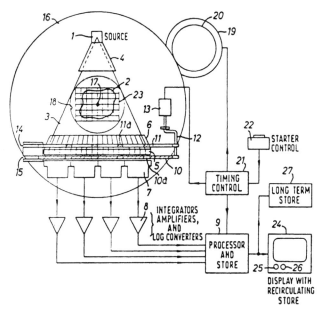

Figure 5.3 A 'rotate-only' CT scanning arrangement incorporating the mechanism for economically coupling more than one detector element to a single photomultiplier. (From Hounsfield (1973–7). Courtesy of the Comptroller of Her Majesty's Stationery Office.)

The major feature of the Hounsfield (1973–7) patent was, however, the method whereby a much smaller number p of photomultiplier tubes was used to record the data from the $n(\gg p)$ detectors. The reason for this was a saving in cost (of tubes) and also to avoid the difficulty of close packing and connecting up some 300 tubes.

The way this was achieved can be seen in figure 5.3. Each of the p photomultipliers were arranged to 'view' a subset n/p of the detectors. Above and below the detectors were oscillating shutters. The shutter 10 between the detectors 5 and the phototubes 7 was opaque to optical radiation with the exception of small apertures, one per tube. The shutter was oscillated to and fro by which means the light entering each phototube was sequentially that from each of the detectors sharing that tube. At the same time a reverse-phase shutter 11 between detectors and source ensured that, at the precise moment that the light from a detector was being read out to the photomultiplier tube, that detector was not detecting a quantum. This was an important arrangement designed to reduce noise. The two reciprocating shutters were activated by a crank mechanism 12 driven by an electrical motor 13. This and the motor 19 which rotated the whole arrangement were synchronously coupled into a timer 21. It was arranged that the shutter oscillated some four times for each orientation of the mechanism in order to sample at a number of times the decay curve of the light characteristic of the scintillator. For computational purposes the data were required at discrete orientations but for mechanical convenience the rotation was continuous. These were reconciled by an interrupting shutter (not shown) which stopped the beam during that part of the rotation when data were not being taken. Thus no confused blurring arose.

Two alternative forms of the reciprocating shutter were also described in this patent, one of which is shown in figure 5.4. 28 is a Perspex cylinder rendered opaque apart from helical transparent grooves 30, one for each photomultiplier tube. The light from the crystals falls continuously on the surface of this cylinder. By rotating the cylinder, each tube sequentially reads out the light decay from the n/p detectors with which it is associated. Just as for the linear reciprocating shutter, there were n/p apertures; so there are n/p helical grooves in this embodiment of the shutter. We might note that the idea of coupling a small number of phototubes to a much larger number of crystals is one now to be found commonly in multicrystal positron emission tomography (PET) scanners (see §6.4).

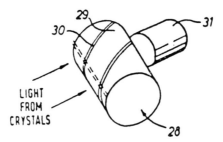

Figure 5.4 Alternative form of coupling a single photomultiplier to a multiplicity of detector elements. (From Hounsfield (1973–7). Courtesy of the Comptroller of Her Majesty's Stationery Office.)

For completeness we note that substantially the same material as in these two UK patents was covered in US patents (Hounsfield 1971–3, 1973–5b, 1974–6a). In divisions of the earlier US patents, Hounsfield (1974–5a,b) incorporates the same material with an extension to describe the use of the water bag for enclosing the head. This feature was incorporated to reduce the discontinuities in radiation absorption at the various angles around the head. The method of water-bag scanning did not survive beyond the earliest head scanners from EMI although other manufacturers patented 'improved' versions (see, for example, Distler *et al* 1975–6).

Hounsfield (1973–5a, 1974–7) also patented a method by which fan-beam tomography could be performed for part of the body enclosed in a water surround. (This patent provided for improved head scanning with a water surround and, although the patent refers to the possibility of scanning the torso, the water bath idea soon fell out of fashion for body scanners.) The arrangement is shown in figure 5.5. The body 1 was enclosed within a rubber bag 6. A Perspex enclosure 5 was shaped substantially parallel to the edges of the X-ray fan beam 3 emanating from tube 4. Water could be pumped into the space between the impervious bag and this frame via pumping arrangements shown at 24 and 25. The X-rays impinged on a collimated detector arrangement at 9. In operation, water was extracted from the space and the patient was inserted. The pump then compressed the bag against the patient and was then removed. The framework 5, together with the source and detectors, rotated during the scan whilst the bag (and of course the patient)

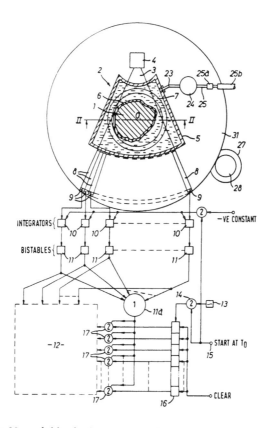

Figure 5.5 Hounsfield's fan-beam CT equipment incorporating a water bath. The system is 'rotate only'. (From Hounsfield (1973–5a).)

remained fixed. A rotary water seal 7 permitted this. The rotation was to have been continuous (unlike in earlier apparatuses) and this patent listed a variety of potential detectors which might be used.

In a further patent (Hounsfield 1974–6b) was described the method by which this scanning could be improved by overcoming the problem of gaps between close-packed detector elements. This was achieved by arranging for the gantry to execute two continuous 360° rotations. At the moment that the second orbit commenced, the whole bank of detectors and collimators were translated or side-stepped by half the detector spacing and the second orbit was accomplished with the detectors in this position. The side step was accomplished by a crank mechanism synchronized into the orbital rotation.

Other patents from Hounsfield and EMI related to CT include descriptions of the special features of the X-ray tube (Hounsfield 1974–6c) and to means for shaping the distribution of intensity across the X-ray beam to improve sampling (Hounsfield 1975–7). They also patented (Hounsfield 1974–6d) a further water-bag body-scanning method by which account could be taken for any lack of centrality of the patient in the field of view. Different data acquisition techniques were patented by Froggatt and Percival (1975–6) and by LeMay (1976–7).

The EMI company had successfully placed 33 brain scanners in UK hospitals by August 1978. There were 22 body scanners installed or on order in the UK by October 1978 (Stocking 1979). However, the number of scanners installed worldwide was an order of magnitude larger. For example 600 EMI brain scanners were installed by this time and some 300 body scanners were installed or on order from EMI (Stocking and Morrison 1978). The first EMI commercial brain scanner was installed, not in the UK, but in the US Mayo Clinic in 1973. The prototype EMI brain scanner had of course preceded this at the Atkinson Morley Hospital, London. The prototype EMI body scanner was installed in Northwick Park Hospital, London.

5.2 A Siemens patent

A UK patent application was filed by Siemens Aktien-Gesellschaft (1975–9) in November 1975 and granted in May 1979. The patent referred to scanning both by the rotate–translate mode (first generation) and by the rotate-only mode (third generation) but did not elaborate experimental details of geometry. The main claim of the patent was a new way to read out rapidly the X-ray measurement of a number of detectors forming a detector bank primarily for fan beam CT. It was envisaged that the detectors might be semiconductor devices (exact types unspecified) and might be ones which have to be kept in a cryostat and operated at a low temperature. Of the order of 256 detectors were envisaged and these were to be coupled via electronic switches, preferably field-effect transistors, to a shift register. The purpose of the arrangement was to allow enough time between reirradiation of the detector bank for the electronics to perform an interrogation of the detector elements in turn. The appropriate time constants were realized by utilizing the internal capacitance of the detectors themselves, additional parallel capacitance and a load resistor. This was not a lengthy patent and it was merely remarked that of the order of 28 000 equations for the 80 by 80 pixel values would be obtained and solved by methods described in the first Hounsfield–EMI patent.

5.3 The Boyd patent

Boyd (1976–8) was granted a patent in February 1978 for a position-sensitive detector arranged to perform three-dimensional CT. The new invention was of a (third-generation) system for rotate-only wide-fan-beam CT. The patent introduced several novel features:

(1) The detector suggested was a position-sensitive multiwire proportional chamber with a focused grid collimator.

(2) The fan-beam projection data were reordered into equivalent parallel beam data sets prior to reconstruction.

(3) A convolution–back-projection method of reconstruction was envisaged.

(4) A monochromatic source was preferable.

(5) Parallel electronics for the multiwire proportional chamber enabled very fast data rates to be used.

Two types of multiwire proportional chamber were proposed. The first was a linear detector with an ionizable gaseous medium such as xenon at atmospheric pressure. Delay line read-out was initially proposed but, to obtain a higher count rate capability, such that imaging could be performed during a breath-holding period, it was suggested that this be replaced by connecting each individual anode wire of the chamber to an amplifier and counter. This enhanced the count rate capability to 10^8 counts s^{-1}. The second improved position-sensitive detector was such that the multiwire proportional chamber was constructed in the form of an arc of a circle and was filled with xenon gas at some 5 atm pressure. This enabled the detector efficiency to remain constant along its length (the linear detector 'sees' less radiation per unit length at its ends than at its centre because of its finite depth) and also improves the overall efficiency and spatial resolution. The latter is achieved because the length of each anode wire can be reduced in that the product of anode wire length and gas pressure should stay roughly constant at 50 atm cm. In general the use of an array consisting of a rotating fan beam and a position-sensitive detector (third-generation scanning) is quicker than those earlier forms of CT scanning employing translation. The equipment with the straight-line detector is shown in figure 5.6.

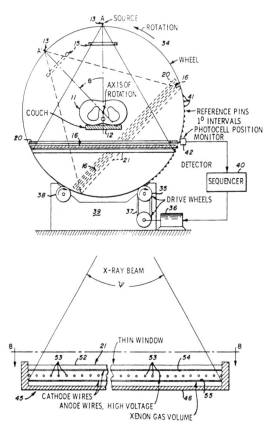

Figure 5.6 One form of apparatus proposed by Boyd. The system employs a wide fan of X-ray beams and is 'rotate only'. The detector is shown in the lower part of the figure, and 16 is a multiwire proportional counter. (From Boyd (1976-8).)

Regarding ray reordering, the patent showed that, if the fan angle was ψ and the apparatus rotates through $180° + \psi$, then equivalent parallel-beam data in the range $0-180°$ may be derived, initially at variable spacing and then restructured to uniform spacing. In Boyd's invention, ψ was $75°$ and hence the apparatus had to rotate through $255°$. Data were captured at each $1°$ interval into 151 projection elements per view. The image was reconstructed on a grid of pixels of size 2.5 mm. The reconstruction method made use of digital convolution functions followed by interpolated back-projection of the convolved projection data.

The preferred source was a pellet of ^{153}Gd arranged in a shuttered housing. Alternatively an X-ray source could be used. The focused grid collimator rejected all but 1% of the scattered radiation. Boyd published much the same material in an earlier US patent (Boyd 1974–6) and in an Australian patent application (Boyd 1978).

5.4 The Lux patent

A substantially different approach to CT was taken by Lux (1978–80) whose patent was granted in December 1980. He recognized the many disadvantages attached to the use of a position-sensitive detector comprising multiple elements. Amongst these were the following.

(1) Each detector generally comprised a small scintillating material requiring its own photomultiplier and amplifier. Matching the gains of these was notoriously difficult.

(2) Such matching led to high operating costs, potential downtime and the need for expert personnel.

(3) The manufacturing costs were very high.

Lux proposed a quite different device in which the position-sensitive detector was a *single* scintillator crystal requiring only one photomultiplier and amplifer. Drift in the gain of this would lead only to a change in brightness of the reconstructed image. The vital position sensitivity was achieved by interposing a rotating aperture between the X-ray source and the body. This aperture was in general opaque to radiation except for a finite number (this number being how many elements were required in the projection—generally 200) of totally transparent windows. The windows were disposed in such a way that one complete rotation of the aperture disc modulated the X-ray beam by a complete linearly independent set of functions which in the embodiment were Walsh–Hadamard functions. That is each position across the narrow fan of X-rays was modulated in a totally different way from all the others. This arrangement enabled the signal from the single detecting element to be related to position within the element by a demodulation process. The aperture required to be rotated (just once) at each orientation of the X-ray source and detector to the patient. Thereafter the data set could be reconstructed in the usual way. The system is shown in figure 5.7 where both the disposition of the aperture and its form are shown. For simplicity the form shown is for a position sensitivity with only four elements; hence there are only four transparent windows in the aperture.

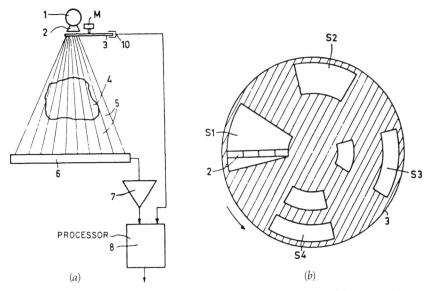

Figure 5.7 The arrangement for X-ray CT scanning proposed by Lux. (*a*) A single detector 6 views the whole of a wide fan of beams 5 from source 1. The detector receives its position sensitivity via an aperture 3 rotated by a motor M. (*b*) One simplified form of the aperture where the beam is modulated into only four elements. The aperture is opaque except for a finite number—in this example four—of open windows. (From Lux (1978–80).)

Several additional points were made.

(1) The aperture did not have to be a binary open or closed aperture. Other modulations were possible including continuous modulation.

(2) If parallel sets of single detector elements were disposed longitudinally along the patient, sets of adjoining slices could be reconstructed from the data with the use of just the single rotating aperture.

(3) Alternatively, using a two-dimensional X-ray fan and a *two*-dimensional modulating aperture, data sets could be obtained from a *single* crystal.

(4) Instead of interposing the modulating aperture between the X-ray source and the patient it could alternatively be disposed between the patient and the detector.

Lux's technique has a certain simplicity in terms of the material requirements. This is offset by considerable complexity in the mathematics of the modulation processes to which here only brief and simplistic attention has been given. Since the sinogram is now a convolution of the true sinogram and the modulating function, the reconstruction problem is even more ill conditioned.

5.5 The Inouye and Mizutani patent

A patent was granted to Inouye and Mizutani (1978–80) in May 1980 for a new form of fan-beam X-ray CT. The specific interest of this patent is that it provided for reconstructing cross-sectional data from a set of projections which did not span the full angular orientation about the patient. The claim was that this was advantageous for the patient in terms of reduced dose and provided for greater comfort in view of the reduced scanning time.

Once again a wide fan beam of radiation was arranged to impinge on a bank of detectors comprising many elements. The source and detector were then rotated about the patient to collect data from only a limited angular range. These data were Fourier transformed into corresponding sets of projections in Fourier space, also of course spanning only a limited angular range. The majority of reconstruction algorithms require the data to span a complete angular range. This patent proposed a method for generating the 'missing' projection data from that collected in the finite span. The principle was that those data which were collected—after transforming to Fourier space—could be expanded in Fourier series. From calculations using the coefficients of these series, the missing data could be arrived at. The details of how to do this were given in the patent. Inouye and Mizutani claimed that the time to reconstruct was shorter than in conventional CT. In addition the apparatus provided for those circumstances when it would not be possible to obtain a complete angular coverage of the object or patient to be imaged. Although the work was presented in terms of calculating from divergent X-ray beam data, the patent drew attention to the possibility of using other probes, for example ultrasound, and also for the limited angle data to be collected in parallel geometry.

5.6 The Swift patent

Swift (1980–1) patented another novel method of performing CT. His method may be easily understood with reference to figure 5.8. A source of X-rays is primarily collimated into a fan which is narrow in the longitudinal direction and has a wide span transversely across the body. Set opposite to this is a single position-*insensitive* detector which may, for example, be a scintillator with one or more photomultiplier tubes and associated detection circuitry. The X-ray beam, however, receives secondary collimation via a rotating chopper 14 which is generally opaque to radiation except for a number of slots 15. As this chopper rotates, it sweeps the slots across the exit of the primary collimation, thus achieving a swept pencil beam of radiation. It is this chopper which provides for the

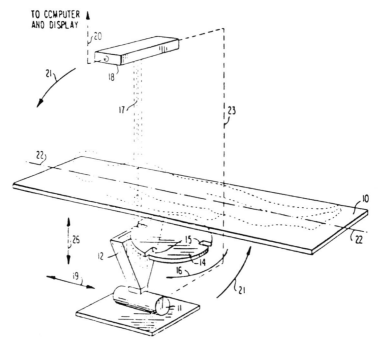

Figure 5.8 A novel form of CT proposed by Swift. There is no multielement detector but instead a single volume detecting element 18. A source 11, collimated 12 to a fan has its exit beam modulated by a rotating disc 14 in which there are several slots 15. The system produces a swept pencil beam. The entire arrangement rotates continuously 21 about the patient on axis 22 generating the projection data. Alternatively the rotation may be dispensed with and instead translation 19 movements made to give a digital radiograph. (From Swift (1980–1). Courtesy of the Comptroller of Her Majesty's Stationery Office.)

positional sensitivity of the detector. Radiation only emerges from the double collimator at the overlap of the common apertures. As the pencil beam reaches one end of the detector, a new pencil 'starts' via the action of a following slit in the rotating aperture.

The manner by which CT may be performed is essentially obvious. The source, collimators and detector rotate about the patient continuously whilst the pencil beam simultaneously sweeps the field. Typically in one embodiment one traverse of the beam took ⅓₀ s and the whole rotation 15 s. Thus some 450 projections are generated. However, the beam may be swept as fast as ¹⁄₁₈₀ s with the rotation accomplished in 5–10 s, giving some 900–1800 projections.

The equipment as shown in the figure is arranged to perform section imaging. However, if the rotation about the patient is dispensed with, whilst retaining the swept beam, a form of digital planar radiology can be performed by laterally translating the equipment relative to the patient. This could be useful for example as a localization 'scout' view prior to CT in order to register the slices. Because the gantry can be angled to any orientation about the patient, such digital imaging is not confined to any particular view and indeed many views may be obtained.

Many advantages over other methods of performing CT are claimed for a system such as this.

(1) Making use of just a single detector, it dispenses with the need to balance multiple detectors to avoid artefacts. It is also much cheaper to manufacture.

(2) The swept beam automatically provides for an excellent scatter rejection; this is a slit method.

(3) The system, whilst essentially a rotate–translate system, nevertheless has very few moving parts and none which requires to be oscillated rapidly.

(4) The use of a large-volume scintillation detector is almost 100% efficient. Hence the system described achieves submillimetre resolution with a dose to the patient of only a tenth to a hundredth of that from other systems.

(5) The method is less complex than others which have proposed reciprocating shutters to give position sensitivity (see, for example, Hounsfield 1973–5b).

(6) The apparatus can have a variable field of view. This may be achieved either by adjusting the physical dimensions of the collimators or by changing the distance between the source and detectors and the centre of the rotation.

(7) In principle the system can be used to generate precisely registered sagittal and coronal slices without breathing movement. This can be achieved by arranging for the width of the fan in the longitudinal direction to be greater and for the beam to impinge on a number of single-element detectors stacked next to each other. A single chopper can sweep the beam across all these detectors simultaneously. Thus the imaging mode is quite different from that in which the patient is sequentially repositioned between slices.

(8) The mechanical scanning arrangement was considered to be an improvement on electronically sweeping the beam in view of the fact that the collimator is nearer the patient.

Swift's technique for developing positional sensitivity is different from that of Lux (whom we recall used a modulation on the X-ray beam) but both are in a similar class of making use of a single efficient detector for simplicity and for the other advantages above. Other forms of collimator can of course be used to give position sensitivity such as a rotating collimator with a helical slit such as proposed for a planar digital radiology system by Arauner (1982). Other patents, too numerous to mention, have made use of different mechanical means of chopping an X-ray beam. Those relevant to section imaging have, however, been covered here.

5.7 The Franke patent

The quest towards faster section imaging led to many proposals. Amongst these was one from Franke (1977–9) which put forward the idea of using multiple X-ray sources disposed at regular angular intervals around the circle encompassing the patient. In the patent, three sources were drawn (figure 5.9). These were able to be deployed in two separate

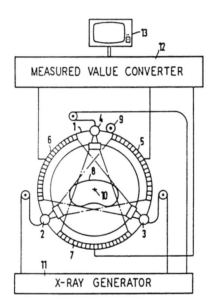

Figure 5.9 The first of the new CT scanning arrangements patented by Franke. Three sources of X-rays 2, 3 and 4 were arranged opposite banks of detectors 5, 6 and 7. The system rotated only 120° to collect full projection data sets. (From Franke (1977–9).)

ways. They could be arranged to produce a fan beam of radiation swathing the patient and incident on three separate arcuate detectors comprising many (typically 240) detector elements; rotation was through 120°, instead of the 360° which would have been necessary with a single source and detector bank. Thus the time for scanning was reduced by two thirds.

In the second arrangement (figure 5.10), a 'rotate—translate mode', the beam from each of the three sources was collimated to a narrower fan incident on only 30 detectors. At each angular orientation, these fans were translated across the field of view via a moving pinhole aperture. To obtain equivalent data sets, the translation was through a distance of eight detector lengths. After each translation the three sources and corresponding detectors were rotated azimuthally by 1°. In this geometry the total angular rotation was only 60°.

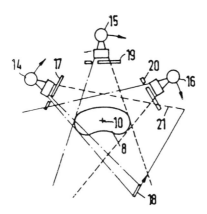

Figure 5.10 The second of Franke's proposals. Again only three sources 14, 15 and 16 are shown with opposing detectors such as 18. The beams were collimated 17, 19 and 20 by lead pinholes and at each orientation the pinholes swept the beam by translation (for example as shown in the direction of the arrow at 18). (From Franke (1977—9).)

5.8 The Wagner patent

Wagner (1975—7) patented a particularly simple method of CT (figure 5.11). A large number N (11 in his drawing) of X-ray tubes were arranged equispaced on the arc of a circle at the centre of which was placed the patient. Each source was collimated to a narrow fan beam which was arranged to impinge on a linear detector bank centred at the opposite end of the diagonal from the source. He arranged that the detector banks were of sufficient length that their ends touched to form an N-sided polygon inscribed to the circle of sources. If all the sources were activated, projection data could be acquired suitable for reconstruction to a section. However, because this would have led to a large scatter contribution it was alternatively arranged for the sources to be activated not quite simultaneously but in quick succession. The data from such an arrangement

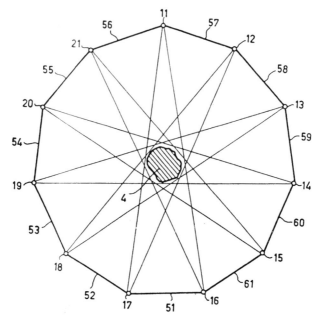

Figure 5.11 Wagner's method for CT. A number of sources—11 are shown—are arranged opposite an equivalent number of linear detector arrays. The patient 4 is encompassed by all beams which are switched on in sequence. The whole set of sources and detector banks executes small angular rotations for each sequential firing until each source has rotated completely to occupy the position of its neighbour at the start of scanning. (From Wagner (1976–9).)

are not numerous enough for reconstruction and this was overcome by making further sets of measurements as above with the source and detector arrangement rotated by successive small angular increments until a full angle of $2\pi/N$ had been traversed. If there were m such angular increments and P detectors per bank, the total number of measurements was mNP, sufficient for reconstruction.

5.9 The Lill patent

Lill (1977–8) patented a method for what is known as fourth-generation CT scanning. By this technique the X-ray beam is collimated to a wide fan (of the order of 40°) impinging on just part of a circular or annular array of discrete detectors which themselves do not rotate with the beam (figure 5.12). The angular extent of the detectors must be at least 180° plus the fan angle so that the recorded data can be reordered into an

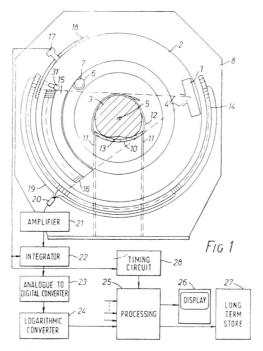

Figure 5.12 Lill's fourth-generation CT scanner. The detectors 14 do not rotate with the source 1 and collimators 16. Their angular extent must be in excess of π by the fan angle. Detectors separated by more than the fan angle can share common data read-out electronics. (From Lill (1977–8).)

equivalent 180° reconstruction using parallel beams. Hence as shown the detectors spanned 220°, were ⅓° in extent and 660 in number. They are shown as 14 in the figure. The advantage of this arrangement over first-, second- and third-generation scanning was that a beam at some fixed displacement from the central or axial beam was not always viewed by the same detector element, which if drifting or out of calibration with the others would have led to circular artefacts.

Lill's scanning arrangement provided for antiscatter collimators 16 which rotated with the beam and were focused at the beam source. By arranging the overlap between the baffles of the collimator and the detector elements to lie at an angle to the detector boundaries. the collimators provided substantially the same exit aperture to each detector, despite rotation.

A further saving in expense and calibration was made by using common electronics 20 for detectors spaced more than 40° apart. Only one of the detectors in any ganged set could of course be detecting radiation at any one time.

5.10 Two more third-generation patents

Many ways were proposed for generating CT images by what became known as third-generation scanning geometry. This was the configuration by which the X-ray beam was collimated to a narrow slit longitudinally and to a wide fan transversely. The X-rays impinged on a detector of at least the same lateral extent as the exit breadth of the fan. In principle the projection data necessary to form a CT image could be obtained by a single rotation of such an assembly.

There were many variations on this theme and two are presented here as examples of different ways of solving the problem. LeMay (1975–8) patented a method using a bank of scintillator detectors. This apparently simple idea whereby rapid CT scans could be achieved suffered from a classic artefact. It proved to be very difficult to ensure that each of the detector elements had the same sensitivity, that is to ensure that the detectors were all matched. Since each element viewed an annulus concentric with the axis of rotation of the scanner, detector imbalance could lead to circular artefacts, bright or dark bands of reconstructed density. LeMay's method overcame this as follows. Instead of the more usual single rotation, the source and detector assembly rotated three times. Prior to commencing each new rotation the detector bank was laterally offset slightly from its previous position. This was achieved by an electromechanical assembly. The result was that essentially the same beam path was then sampled by three detector elements. If they registered different values, this could only be due to drift and lack of matching. These unwanted properties could thus be calibrated away.

The problem of matching detectors led several workers to use an alternative which was considerably more stable. Taylor (1976–7) arranged for a detector to be fabricated from a plurality of cells containing high-pressure xenon. Once again the X-ray fan was directed in a swathe towards the detector and the whole assembly rotated one revolution to form the data. Xenon detectors, as noted previously (see §5.3), are considerably more stable and the problem of drift in sensitivity was largely eliminated. Xenon gas at 5–30 atm was used. The tank comprised some 300 cells arranged by introducing separators into a single tank. The separators were very thin (approximately 0.25 mm) and of tungsten, tantalum, molybdenum or platinum. It was necessary for the charge built up in a cell during one integration period in response to the radiation to be removed before radiation was received in the cell from another path in the body. To this end the radiation beam was interrupted periodically by either mechanical or electronic means.

5.11 Compound computed-tomography scanners

In principle, provided that a plurality of beams can be arranged to traverse a section, that section can be reconstructed by mathematical techniques.

The experimental methods for achieving the projection data are, as we have seen, generally classified into one of four groups or generations. The first two generations of scanners required translational movements as well as rotational motion, generally for a large number of angular orientations about the patient. Apparatus for performing these rectilinear scans was massive and required accelerating, decelerating and reversing direction of the source—detector sets hundreds of times per scan. This of necessity led to long scanning times as well as to very complex engineering. Cox and Snyder (1975–6) proposed a novel scanning method to overcome these limitations. The main feature of their invention was the ability of the source to pivot about an axis normal to the scan plane and through the source point. Correspondingly the detector could also pivot on a radius connected to the X-ray source point. That is the source and detector had individual pivotal movements coupled such that the source and detector were always facing each other. These movements (which were usually absent in other scanners) were in addition to the orbital rotary movement of the whole assembly about the central axis of the system. The detector comprised many elements but the equipment could be used in many ways as follows.

If the X-ray beam were collimated in the plane of the scan to only a small sector impinging on just one or a few detector elements, it was necessary for the whole assembly to execute many orbits. At the start of each orbit the source and detector were incrementally pivoted about the axis at the source, this motion being locked for each orbit. For the first orbit the source and detector were maximally offset in one direction from the line through the orbital axis and for the last orbit they were maximally offset in the other direction. In between they took up all the other sector paths which in total made up a wide fan. It can therefore be appreciated that no rapid accelerations or decelerations and changes in direction of movement for heavy equipment were required. The gantry rotated at 15 rev min^{-1} and the source and detector were pivoted 1° as each orbit was completed. The total pivot of the source and detector was 40°.

In essence the limited angular coverage of the beam at each orbit was traded against a multiplicity of orbits. This compound motion was easy to implement and capable of many variations. It could, for example, be reduced to the 'classic' third-generation scanning mode if the opening angle of the X-ray collimators were arranged to provide a full-coverage wide fan and the detector were a position-sensitive bank, for example, of scintillators. In this form, only a single orbit was required. Alternatively intermediate multiorbit compound scanning arrangements may be envisaged. The data capture electronics for compound scanners of this type has been described in detail by Brunnett (1975–7). The equipment could be used to provide CT scans with a variety of spatial resolutions. For example, if a wide swathe beam were used and nominally a single orbit, the resolution could be improved by adding in the data from a second orbit with the source–detector assembly pivoted by just 1°. Both of the patents described in this section were assigned to the Picker Corporation.

5.12 Features from other patents

There are an enormous number of patents and inventions for CT and some selection of key ideas is necessary if the account is not to become too indigestible. The approach here has been somewhat selective and not all patents have been reviewed. The philosophy has to be rather different from what was possible when reviewing pre-War patents! In this section we group together ideas which might be thought of as variations on a theme.

Kalender *et al* (1980–2) patented a method for quantitating the time dependence of reconstructed pixel values in a small part of a CT image. The method comprised initially performing conventional CT using a fan beam impinging on a detector with 256 elements. This having been achieved the region of interest in the image can be selected where time-dependent data are required. The source and detector orientation together with the patient couch are locked in this position. The wide X-ray fan is subsequently collimated across the usually wide angle such that only those detectors in line with the region of interest are exposed. From this arrangement the data may be used to reconstruct time-dependent pixel values for just this limited sector field of view.

The idea that apparatus for CT could be used for digital radiography was noted by many workers. Specifically the idea was patented by Liebetruth (1977–9). The so-called scout view was generated to enable a precise positioning of the patient relative to the gantry for slice imaging. A similar idea had been described also by Distler (1975–6).

Richey *et al* (1977–9) noted the need to provide a means of improved X-ray collimation when multiple detectors were used in fan-beam geometry and patented a suitable device. The implication was that the moving-shutter devices of the Hounsfield type were cumbersome and complicated. The new collimators proposed enabled a variable field of view.

Wagner (1976–9) developed new computational techniques with improved interpolation for CT scans with faster reconstruction times.

A very interesting development of an adjunct for tomographic imaging was patented by Froggatt (1977–8). This concerned the well known problem when delivering radiotherapy via rotation of the source relative to the patient—that the patient (and thus the tumour being treated) may move during the therapeutic exposure by small but significant distances. Froggatt provided a detector array, based on the use of scintillators or a fluoroscopic screen, which recorded a transmission projection during the therapy. Such projections were able to be computed repeatedly for each orientation of the beam relative to the patient. Prior to this a diagnostic CT scan of the patient in the same position had been made. From this a computer calculated the projection data for all angles to be visited by the therapy beam. The projections during treatment with the gantry at some or all orientations were compared with those calculated from the therapy scan. From the comparison the required beam shift was executed to re-register the treatment beam with the tumour. The patent goes on to point out that two-dimensional projection data may be likewise calculated and used to make out-of-plane adjustments. This problem described 10 years ago is still considered by some to be largely unaddressed in practical terms.

Froggatt (1976–8) was concerned with the problem of scatter into the elements of a detector bank in that form of third-generation CT scanning when a wide fan beam impinged on a detector bank but when at a specific orientation the focal point of the X-ray beam was translated a small distance in the tube by electronic steering. Under these circumstances, detector collimation was inappropriate because each detector element had to be able to receive radiation from a number of directions.

5.13 Charged-particle computed tomography

In §4.6.1 we saw how Cormack foresaw the possibility of proton CT and how this was achieved with Koehler. Workers at the Donner Laboratory of the University of California were also working towards transverse axial scanning with high-energy charged particles (Crowe *et al* 1975, Budinger *et al* 1975, Budinger 1987). Their aim was to achieve comparable spatial resolution and electron density resolution with just a fraction of the dose needed for X-ray CT. This was achieved (figure 5.13) by arranging for a beam of 900 MeV alpha particles from a rectangular slit, thin in the longitudinal direction and spanning the patient in the axial direction, to be detected by a multiwire chamber and a range-counter telescope. Projections were recorded by rotating the patient on a chair in the beam line to just 32 angular orientations and CT data were reconstructed by iterative and Fourier transform algorithms. Images of the brain were achieved with the same density resolution as an EMI scan but with a dose of only 30 mrad instead of 1600 mrad for the latter. The experimental technique used in 1975 was a prototype. Like many other early explorations the interrogating beam was kept static whilst the patient was rotated. It is probably fair to say that in contrast with the widespread development of X-ray CT, charged-particle CT has not proved so popular especially as it requires a linear accelerator to be available.

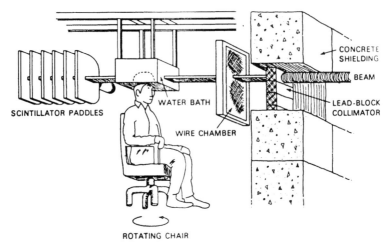

Figure 5.13 Tomography at the Donner Laboratory using accelerated charged particles. (From Budinger (1987).)

5.14 Summary

In this chapter a different approach has been taken to looking at the modern history of X-ray CT through the primary source data of patents. The technology developed very rapidly between around 1968 and the end of the 1970s. Of necessity not all patents even in this limited period have been covered but the major developments have been noted. The order of presentation has not been carefully sequential because these developments were all in a short space of time and have overlapped in time. To put them in (say) the order of patent application (or granting) would not significantly enhance the story.

The role played by the EMI Company in implementing CT was pioneering. Very soon a number of other companies were marketing commercial CT scanners and the affiliations of the patent holders can be found on the front pages of the patents concerned. Models came and went fast and, to avoid any misunderstanding, this account has refrained from describing specific machines and their features, preferring to document the physics underlying them. Hounsfield's scanners included first-, second- and third-generation designs. Third-generation scanners were state of the art for a number of years and there were many improvements to the basic design. For example there was the improved stability of using proportional counters (Boyd, Taylor), the use of multisources to reduce the scan time (Franke, Wagner), reduced angular coverage (Inouye and Mizutani) and the method of LeMay which overcame detector balancing by introducing a small source nutation. Cox and Snyder introduced compound scanning to avoid problems of accelerating massive machinery. Balancing the response of individual detectors proved to be a major problem with third-generation machines and to overcome this we have seen how two workers (Swift, Lux) invented beam choppers so that a single detector could be used with position sensitivity. Eventually fourth-generation machines became the state of the art (for example, Lill) and the problems of ring artefacts went away.

The use of fast-data-capture electronics and digital computers for reconstructing both accelerated the progress of CT and enabled quantitative maps of body attenuation to be formed. In all respects the images obtained were stunning to the point that those whose radiological experience does not pre-date the routine use of CT would have little time for classical tomography.

The attempt in this chapter has been to distil the essence of the development reported in patents. Patents, of course, differ very much in their descriptive style from papers; they use a somewhat legalistic style for the purposes of protecting all aspects of the claim and do not always come to the point quickly. In each patent the new science is presented at least twice; first in a descriptive form and then once again as a list of claims. To a large extent this style has been avoided here and the key points of each new idea have been summarized more in the form expected by the readers of scientific papers.

CT is still a major part of the diagnostic armamentarium despite the emergence of new imaging techniques such as nuclear magnetic resonance and the great improvements in imaging by diagnostic ultrasound. It has been reasonably established that scanners are now only quantum limited; so further grand improvements would not be expected. Scanners are everywhere and the hardware cost of installing new machines is often not such an important consideration as the running costs. As this chapter began, it concludes by acknowledging the immense importance of CT. The origins of how this has come about and in so short a time have been discussed.

The implications for health care of the development of CT scanning are many and varied. These include the proper consideration of capital and revenue costs, the effect of these burdens on health care finance and the balance with the provision of other services. Stocking and Morrison (1978) studied such questions in some detail and particularly the relative roles of philanthropic versus National Health Service funding in the UK and the cost–benefit of making use of CT data. They wrote, 'With its strong desire to believe in miracles, whether provided by a new procedure or a new treatment, the public welcomes any new technology with open arms. . . . It is not difficult to foresee that some technologies in the future will be so costly in relation to benefits that society will be forced to renounce them.' This was over 10 years ago. Whilst CT scanners are still not cheap, they have become widely available, they are largely regarded as routinely justifiable and a considerable body of evidence has accrued as to their scientific and medical usefulness. To some extent the same questioning is now diverted to the newer imaging modality of nuclear magnetic resonance.

Chapter 6

Historical Emission Tomography

6.1 Scanning, stereo imaging and pseudotomography

In this chapter, attention is turned towards emission tomography. Emission imaging is much more modern than X-ray transmission imaging since the external imaging of photons from radionuclides within the body did not begin seriously until the late 1940s. Boag and Smithers (1967) noted that the Royal Marsden Hospital was the first in Britain to start clinical investigations with artificial radioactive isotopes in 1948. Imaging of the brain, for example, was a pressing concern but isotope methods took time to become established. Commenting on the use of isotopes in neuroradiology, Fischgold and Bull (1967) wrote that, at the time of the Fourth Symposium Neuroradiologicum in London in 1955, only two papers were presented, one on the use of ^{42}K (Kramer, Burton and Trott, London) and the other on ^{74}As positron imaging (Brownell and Sweet, Massachusetts): 'The sessions devoted to isotopes left one with no firm impressions as to their value in neuroradiological diagnosis.' By the 1961 Rome Symposium, isotopes for brain imaging were firmly established. An intriguing history of the development of nuclear medicine has been provided by Brucer (1978). The first nuclear medicine imaging involved simple counting experiments with no attempt to obtain tomographic localization.

In speaking of the 'history' and origins of emission tomography (McRae and Anger 1974) it is necessary to concentrate on the decade or so beginning with the work of Kuhl in 1963 (see §4.3). The period between 1950 and this time was firmly the era of planar scanning and, although focused-collimator scanning possessed mildly tomographic features, it cannot really be regarded as a firm attempt to localize activity in a patient on a point-by-point basis. Hence the detailed developments of scanning technology will not be treated here, with the exception of one machine due to Cassen. There are many reviews of scanning including that by McAfee *et al* (1966). An early attempt to make scanning more tomographic was made by Hart (1965) who made use of coincidence detection for those radionuclides emitting two gammas in cascade (see also §6.2.9). Similar suggestions were made by Arimizu (1972) who also experimented with other methods of longitudinal-section imaging similar to those below. Apart from these brief reports on cascade and coincidence imaging, scanning was largely non-tomographic. After Kuhl's work, the first significant moves towards tomography were due to Harper and Anger (born 1920) (see §§6.2.1 and 6.2.2).

Before moving on to this, however, it is worth noting that a strongly focusing longitudinal imager was constructed by Cassen. Cassen is often quoted as the father of rectilinear scanning although there was some earlier work by Ziedses des Plantes (1950a,b) (see §1.4). Cassen (1968) showed results from a rectilinear scanner that possessed a rather remarkable detector assembly (figures 6.1 and 6.2). The detector comprised a multicrystal layer of broken pieces of sodium iodide immersed in arochlor oil. The crystal matrix, of mass 23 kg, was viewed by seven photomultiplier tubes. A collimator was provided comprising 2200 holes in the

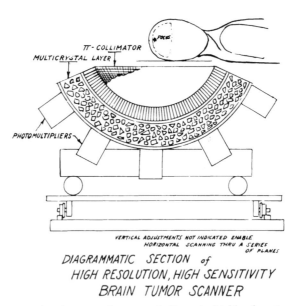

DIAGRAMMATIC SECTION of
HIGH RESOLUTION, HIGH SENSITIVITY
BRAIN TUMOR SCANNER

Figure 6.1 A schematic diagram of Cassen's highly focusing section imager. (From Cassen (1968) in *Fundamental Problems in Scanning* ed A Gottschalk and R N Beck. Courtesy of Charles C Thomas, Publisher, Springfield, Illinois.)

Figure 6.2 A photograph of the collimator in Cassen's scanner. The collimator holes are clearly visible and beneath can be seen some of the photomultiplier tubes. (From Cassen (1968) in *Fundamental Problems in Scanning* ed A Gottschalk and R N Beck. Courtesy of Charles C Thomas, Publisher, Springfield, Illinois.)

form of a spherical cap, strongly focused to 18 cm and subtending an angle of π (that is one quarter of a total sphere). Broken crystals were used because at that time it was not possible to grow large crystals cheaply.

The rationale for the development was to increase the collection efficiency of a scanning system. However, when the isoresponse profiles for ^{203}Hg and ^{99}Tc^m were examined, it was realized that the wide-angle viewing geometry resulted in a very shallow depth of focus. Laminar-section tomography had resulted somewhat by accident.

Not surprisingly several workers tried to adapt stereographic techniques to nuclear medicine. Amongst the first were the attempts by Levy and Okezie (1963) who reported that with phantoms 'near perfect representations were achievable in a single stereo picture'. Successful stereo imaging of the thyroid gland was reported. The very first attempts by Kuhl at differential imaging were based on stereoscopic imaging with two collimated detectors (see §4.3). He was less enthusiastic and soon changed to other section-imaging techniques. Some success came later with the use of the Anger camera to form stereoscopic views (Charkes and Williams 1974). The method did not, however, achieve wide popularity.

Table 6.1 Early US patents for emission tomography.

Date		Author	Number
Registered	Granted		
8 November 1968	12 October 1971	Walker	3612865
17 March 1970	6 July 1971	Brill *et al*	3591806
13 April 1970	15 August 1972	Muehllehner	3684886
28 September 1970	30 January 1973	McAfee *et al*	3714429
26 June 1971	8 January 1974	Miraldi	3784820
24 August 1973	3 December 1974	Muehllehner	3852603

Another pseudotomographic method was that of rotational cinescinti-photography. The patient was imaged at 36 sequential orientations by standing on a turntable in front of a gamma camera (Handmaker *et al* 1974). Each 15 s exposure was captured on 16 mm movie film. The film was then spliced to form a continuous loop and projected, resulting in a moving display with apparent depth information. The technique cannot be regarded as true tomography but there is no doubt the impression of depth was generated and a modern analogue of the method is still some-times used whereby gamma camera two-dimensional projections are snaked rapidly before the eye on a visual display unit. The effect is, for example, particularly dramatic when viewing uptake in the whole skeleton. The major landmarks in emission tomography are summarized in tables 6.1 and 6.2 and these will now be discussed in turn.

Table 6.2 Some of the principal early landmarks in section imaging of radionuclide distributions.

Year	Worker; invention
	Transverse-section imaging
1963	Kuhl; section imaging with scanning single detectors (for later 'Marks' of the Kuhl scanner, see table 4.1)
1965	Harper; three-dimensional display from gamma camera data
1965	Hart; coincidence tomography
1967	Anger; SPET via rotating patient viewed by stationary gamma camera; analogue reconstruction

Table 6.2 *(continued)*

Year	Worker; invention
1968–74	Budinger; SPECT via rotating patient and static camera with long-hole collimator
1969	Brill *et al*; multicrystal transaxial scanner (Vanderbilt Mark 1 Tomoscanner)
1970	Monahan and Powell; coincidence tomography
1971	Muehllehner and Wetzel; SPECT via rotating patient and static camera
1972	Myers; Aberdeen implementation of Kuhl-type and Anger-multiplane type scanning
1972	Arimizu; coincidence tomography
1974–6	Keyes; Humongotron
1976	Pickens; Vanderbilt University Mark 2 Tomoscanner
1977	The beginnings of commercially available SPECT
	Focal-plane tomography (longitudinal-section imaging)
1963	Kuhl; focal-plane tomography with scanning detectors, stereoscopic imaging
1965	Cassen; large-solid-angle scanning detectors
1966	Anger; multiplane tomographic scanner
1967	Follett; rotating parallel hole collimator
1968	Walker; patent for rotating parallel-hole (angled-hole) laminography
1969	McAfee; scanning angled-hole collimated detector
1969	Miraldi and Di Chiro; Tomoscanner
1970	Muehllehner; rotating parallel-hole (angled-hole) collimator and compound-motion laminography
1972	Freedman; rotating parallel-hole (angled-hole) collimator, reconstruction by computer
1972	Rudin; rotating magnifying collimator
1972	Cottrall; rotating parallel-hole (angled-hole) collimator
	Positron tomography
1951	Wrenn; dual-detector imaging
1953	Brownell and Sweet; dual-detector imaging
1960	Higinbotham and Rankowitz; the 'hair drier' at Brookhaven
1961	Robertson; first ring system (32 crystals) at Brookhaven (the 'head shrinker')
1963	Anger; work with a gamma camera
1968	Brownell; first multicrystal area detector at MGH
1976	Muehllehner; double-gamma-camera system

6.2 Transaxial emission tomography

6.2.1 Harper and three-dimensional display of gamma camera data

Harper was among the first to point out that the images from a gamma camera at different angles around a patient could be used to generate a three-dimensional distribution (Harper *et al* 1965, Harper 1968). He proposed that the images could be recorded on 35 mm film and arranged in a circle as shown in figure 6.3 with a central source of illumination. The shadows of the film were directed to a ring paraboloid zone mirror and subsequently deflected to superimpose in a viewing region. The mirrors

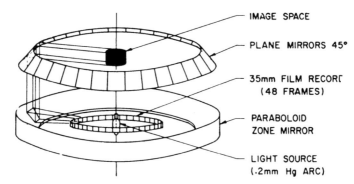

IMAGE SPACE

PLANE MIRRORS 45°

35mm FILM RECORC (48 FRAMES)

PARABOLOID ZONE MIRROR

LIGHT SOURCE (.2mm Hg ARC)

Figure 6.3 The arrangement proposed by Harper for producing a three-dimensional viewing effect from data captured by a rotating gamma camera. The camera images were transferred to film; the strip of film was arranged in a circle as shown and the images were superposed at a translucent screen. (From Harper (1968) in *Fundamental Problems in Scanning* ed A Gottschalk and R N Beck. Courtesy of Charles C Thomas, Publisher, Springfield, Illinois.)

corrected for the divergence of the camera data and in the viewing region a translucent screen rotated rapidly to allow the eye to synthesize the full three-dimensional display. This was an early example of the attempt to see many body sections simultaneously in contrast with the Kuhl single-section imager. In 1968 Harper wrote, 'Although the Anger camera will probably never completely replace the moving detector system, especially when fine resolution is critical, it clearly opens new avenues.' Time has proved the role of the Anger camera in emission tomography has been greater than he imagined, that rectilinear scanning has virtually ceased but that single-plane section imaging machines (with moving detectors) still have their supporters (see §6.2.10).

6.2.2 Single-photon camera-based emission tomography by Anger

A more practical attempt to form cross-sectional images from data collected on a gamma camera (that is single-photon emission tomography

(SPET)) was made by Anger as early as 1967 and reported (Anger *et al* 1967) in June at the Fourteenth Meeting of the Society of Nuclear Medicine. The patient was rotated in front of the static camera, continuously, and single lines of camera data (projection strips), corresponding to emission from some particular slice, were back-projected onto a rotating film, turning in synchrony with the patient, in the scope camera. In order to achieve back-projection in analogue form an array of glass rods was placed in front of the lens of the camera to transform the scintillating dots (projection elements) into lines. The technique of using a fixed camera and rotating the patient was copied for some years before it became possible to rotate a gamma camera (see also §6.2.5). Anger (1974a) himself remarked that this method of achieving single-photon emission computed tomography (SPECT) by analogue means was not really very successful. The method was rather cumbersome and it was not really until the reconstruction process could be achieved on a digital computer and methods for rotating the camera instead of the patient were engineered that the technique became practicable as a routine investigative tool.

6.2.3 Single-photon emission computed tomography

Anger's work used analogue techniques to build up section images. The first reported use of a computer for this purpose was by Muehllehner and Wetzel (1971) who reconstructed projection camera data from just eight views using a back-projection technique. This put the 'C' in SPECT. At that time the technology to rotate a heavy gamma camera was still not available and the camera was kept static whilst the patient rotated in a chair in front of it. In view of the small number of contributing views the section images were subject to artefacts. Attempts to include the effects of photon attenuation were considered. The reconstruction was performed on an IBM 360/30 computer and a matrix 40 by 40 pixels in size was formed. Iterative reconstruction techniques were developed and it took approximately 45 s per slice (35 iterations) using a basic assembly language. Muehllehner and Wetzel pointed out that a significant advantage of their section imaging over other developments was that only double the number of 'conventional' planar camera views were required and with these captured and viewed a clinical decision could be taken whether to proceed or not to reconstruct transverse sections. They also made it clear that they were not at that time claiming that SPECT was clinically useful—indeed they regarded this as generally unproven—but that this question would in time be answered as has indeed been the case. Their clinical efforts were largely confined to the head and they noted the difficulty in extending to the liver in view of photon attenuation in the abdomen.

It was in fact some years earlier that work had started in this direction. Muehllehner first presented the results at the 1968 Society of Nuclear Medicine Annual Meeting. At that time he was working for Nuclear Chicago Corporation and publishing material was not a priority. This explains the 3 year time span between this oral presentation and the ensuing paper. In recollecting this period, Muehllehner (1988) wrote, 'Since this was one of my first papers in the field of nuclear medicine, I clearly recall the experience of having the famous Dr Kuhl approach me to express his interest in hearing my presentation.' Muehllehner was also instrumental in suggesting the need to modify the gamma camera to avoid the problem of accommodating the patient's shoulders. The later work was pushed forward by Jaszczak in the mid-1970s (see, for example, Jaszczak *et al* 1977).

6.2.4 *Single-photon emission computed tomography at the Donner Laboratory*

A number of workers developed techniques for imaging transaxial sections from gamma-camera data, beginning with Anger himself in 1967. In continuing this development the work at the Donner Laboratory and Lawrence Berkeley Laboratory at the University of California at Berkeley was pre-eminent in the early 1970s. Notable among those responsible for progress at this centre were Budinger and Gullberg.

At that time the only way to gain access to the emission projection data suitable for reconstruction into sectional images was to rotate the patient as accurately as possible in front of a stationary gamma camera. The unique feature of the Donner Laboratory implementation was to incorporate the use of a special parallel-hole collimator for the camera with long holes. This was used to enhance the collimation for the camera which necessarily had to be some distance from the head of a subject in order not to collide with the shoulders as the patient rotated. (These were the days before the idea of either tilting the camera or using a cut-away camera head to overcome this problem became possible.) Necessarily the concept of rotating the patient relative to the camera rather than the other way round, as we know it today, led to much time and effort being expended in ways to immobilize the patient. The situation was far from ideal and yet the images obtained showed much detail. It was not until the late 1970s that commercial gamma-camera manufacturers overcame the difficulties of accurately rotating heavy gamma cameras and SPECT as we know it today became commonplace with the patient lying down on a bed around which the camera rotates.

A further difference between modern SPECT and SPECT at the Donner Laboratory in the developmental years was the small number of contributing camera projections or views. Early images (Budinger 1974) were reconstructed from as few as 18 contributing views.

Perhaps in marked contrast with these hardware limitations, much of the reputation of the Donner Laboratory work centred on the exhaustive development of reconstruction algorithms and in particular iterative techniques. Even a review as early as 1974 (Budinger 1974) showed that the development of reconstruction techniques had included all those in common usage today as well as a large number which are still not implemented in commercial equipment. The convolution and back-projection method was firmly established—as well as the equivalent two-dimensional Fourier method—and in addition to the algebraic reconstruction method and the simultaneous iterative reconstruction method due to others (notably Herman (1972) and Gilbert (1972a)), Budinger and Gullberg developed the iterative least-squares technique which in trials with phantoms they found of superior accuracy, generated less artefacts and gave good resolution. Unfortunately they noted what is still largely true today that convolution methods were much faster (their particular implementation was 80 times faster) than the iterative least-squares technique and sadly such iterative methods although available since their work have not really been followed up. The group's interest in this branch of imaging theory goes back as far as 1968 and has been encapsulated in a much used and widely distributed software package known as the RECLBL library (Huesman *et al* 1977). In principle the iterative reconstruction techniques, which all utilize reprojections or pseudoprojections, can account for photon attenuation and even scatter if suitable measurements of linear attenuation coefficient and assumptions about scatter distributions are incorporated. These were not part of the early development but were problems later tackled at this (and some other) laboratories. Since these early days, algorithms for quantitative SPECT have become increasingly more sophisticated and the literature of reconstruction from projections is enormous and well beyond the scope of our collection of early work on the origins of emission tomography. Contributing to the physics of SPECT development was just one of the many important achievements to come out of the Donner Laboratory (Budinger 1987) among which not least was of course the invention of the Anger camera itself. More modern imaging developments include high-resolution ring positron tomography and transmission brain imaging with helium and neon ions (Crowe *et al* 1975).

6.2.5 *Keyes' Humongotron*

The first really significant equipment to achieve SPECT in the form in which we know it today was built by Keyes *et al* (1977). A gamma camera was attached to a rotating gantry via a counterbalanced C arm and was able to rotate about the patient who was cantilevered on a couch into the field of view. The equipment was christened 'the Humongotron'.

The radius of rotation of the camera could be varied by motor drives and studies were performed with 30, 60 and 120 contributing views depending on the balance between the time available and the need to avoid artefact production. Digital data were collected and reconstructed by computer. The machine was used largely for imaging the brain. An attempt was also made to investigate the performance of the system with the camera in continuous rotation. The equipment may have looked a little ungainly by the standards of some modern gantries (figure 6.4) but in most respects it achieved the same and may rightly be regarded as one of the principal precursors of modern SPECT based on the use of a rotating gamma camera.

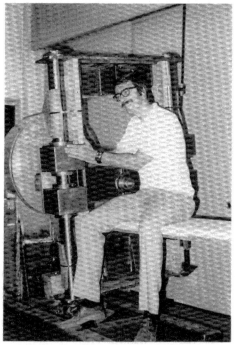

Figure 6.4 John Keyes with the Humongotron, the first machine to rotate a gamma camera successfully about a patient. (Courtesy of J Keyes.)

The Humongotron was the culmination of many years' efforts by Keyes and his colleagues towards single-photon tomography (Keyes 1988). He became interested in the subject of using a gamma camera in 1970 when a junior faculty member at the University of Michigan. With a graduate student D Strange, he developed a small computer program operating within the confines of an 8 bit minicomputer (Nuclear Data 50/50 Med Computer System) to do simple back-projection from single gamma-camera profiles. The system was tested by borrowing a rotary milling machine table from the student shop and mounting on top of it a home-made phantom which was indexed by hand. Keyes (1988) wrote, 'Much to our great amazement, we discovered it actually made recognizable pictures.'

Shortly thereafter Keyes moved to the University of Rochester where with D Kay and a grant from the National Institute of Health he developed filtered back-projection on an IBM mainframe. The method was adapted by D Leys for another Medical Data Systems Computer and using this software and a home-made rotating table and chair, patient brain studies were performed.

Late in 1974, Keyes returned to the University of Michigan to work on a full tomographic imaging device based on the use of an Anger camera. After several false starts, one of the students on the project came in one day to say that the University Surplus Property Shop had a discarded obsolete caesium therapy machine which would do nicely for the rotating gantry. This was acquired for the nominal sum of US$100 and a Rumanian refugee engineer by the name of Nikki Orlandea designed the modification which would allow the attachment of the gamma camera and rotate it about a cantilevered patient. The machine was built in the University Physics Shop in the summer and fall of 1975. It was during this time that it acquired its name. Dr Les Rogers walked into the shop one day and commented that the machine was going to be 'humongous'. It was promptly dubbed with its new name.

Keyes remembers that the name was not entirely approved of. When he originally submitted the manuscript of the paper describing the machine, the editor and editorial board of the *Journal of Nuclear Medicine* felt the name was unscientific and inappropriate for inclusion in the journal. Keyes wrote back, pointing out that, if Professor Mallard in Aberdeen (see §6.2.7) could call his machine the Aberdeen Section Scanner (or ASS for short), he could call his machine the Humongotron. Not a further word was heard from the editorial office and the paper was published with the name unaltered.

6.2.6 The Vanderbilt University scanners

6.2.6.1 The Vanderbilt University multicrystal transaxial Tomoscanner (Brill et al). A novel approach to transaxial tomography was taken by a group at Vanderbilt University (Patton *et al* 1969, 1972, 1974). The machine was patented by Brill *et al* (1970–1). The device may best be understood by reference to figure 6.5(*a*). Two circular rings 1 were arranged separated by four rigid equally spaced rods 5. Between reach rod were arranged two tubes 3, supported at each end by the rings 1 and capable of rotation in ball races. To each tube was attached rigidly a collimated detector,

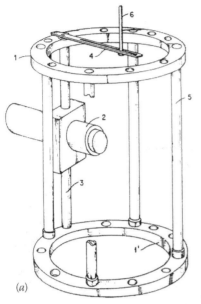

(b)

(c)

(e)

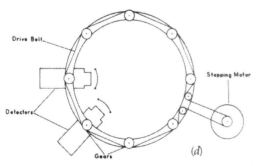

CONSTANT ANGULAR VELOCITY DRIVE

Drive Belt

Stepping Motor

Detectors

Gears

(d)

comprising a scintillation crystal fitted with a 19-hole converging collimator (designed by Beck). Also rigidly attached to each tube were slotted levers 4. It may then be understood that movement of the levers caused the detectors to rotate on their supporting tubes. In the figure, only one of the eight assemblies is shown. The slotted levers were all arranged to intersect at a common spindle 6. This spindle was caused to execute a raster scanning motion (originally by a cannibalized rectilinear scanner, later by a special-purpose $x-y$ driving assembly). In this way all eight detectors were forced to view a single point in space and this point executed the raster scanning movement.

The whole arrangement was assembled horizontally so that, for example, the head of a patient could snugly fit within the scanned field of view. This is shown in figures 6.5(b) and 6.5(c). The technique was based on arranging for the reinforcing of signals from some specific point in space at the expense of others to achieve a tomographic effect. A second form of scan control, which was termed the constant-velocity method is illustrated in figures 6.5(d) and 6.5(e). The detectors are all constrained to execute a rocking motion on their axis via coupling to a drive belt and a stepper motor. The detectors each sweep an angle of 68°. In this mode, reconstruction methods become practical and several were tried including iterative techniques as well as simple summation which was equivalent to simple back-projection. In this mode the outputs from the detectors were stored separately for this purpose. A PDP9 computer provided the control. This method for making single section scans using multiple detectors contrasts strongly with camera-based techniques.

6.2.6.2 The Vanderbilt University Orthoscanner (Pickens). Experience gained with the Vanderbilt University Mark 1 scanner led to a second quite distinct machine being engineered at the same university, christened the Orthoscanner. Describing it, Pickens (1977) wrote, 'The Orthoscanner may be considered as a sort of second-generation device which incorporates concepts and ideas that were produced as a result of experience

---

Figure 6.5 (a) The patented design for the Vanderbilt University multicrystal Tomoscanner; see text for description. (From Brill *et al* (1970–1).) (b) An end view of the Vanderbilt University Tomoscanner showing the rectilinear scanning mechanism which rastered a pin connected to the focal points of eight collimated detectors. (From Patton *et al* (1974).) (c) A side view of the Vanderbilt University Tomoscanner showing how the head of a patient was inserted into the region between the detectors. The rectilinear scanning arrangement is to the left. The whole apparatus was moderately portable as can be seen from how it can be brought up to a patient couch. (From Patton *et al* (1974).) (d) The modified Vanderbilt University scanner (schematic). (From Patton *et al* (1974).) (e) An end-on view of the modified Vanderbilt University Tomoscanner. The eight detectors can be clearly seen. (From Patton *et al* (1974).)

with the Vanderbilt University Tomoscanner of Patton *et al.*' The machine was the idea of Jim Patton and Paul King. It was largely designed and built by David Pickens in the laboratories of A Bertrand Brill (King *et al* 1971, Pickens *et al* 1978).

The aim was to overcome the inherent difficulty with the Tomoscanner of obtaining a spatially invariant point spread function. The Orthoscanner not only removed this problem (which then enabled space-invariant post-processing of images) but also possessed increased sensitivity. This was achieved by the following geometrical arrangement of single-crystal collimated detectors. The basic component of the scanner was an orthogonal triad of detectors, arranged as the name suggests on three orthogonal axes and focusing at a common point. Four such triads were constructed (figure 6.6) and attached to a rigid frame. Eight of the detectors lay in the $x-y$ plane with the other four orthogonal to this plane. The triads possessed collimators with 4.5 in focal length and were set in the frame at 4.5 in spacing. The rigid arrangement of 12 detectors was arranged to scan in the $x-y$ plane in a rectilinear fashion (figure 6.7) such that one quarter of the square field (which was 9 in by 9 in) was scanned by the common focal point of each triad (and thus, there being four triads, the whole square field was covered). Similar movements were arranged in the z direction; so in all a cube of side 9 in was scanned by this sensitive point method. This was large enough to contain the head which was the clinical region of interest for this group.

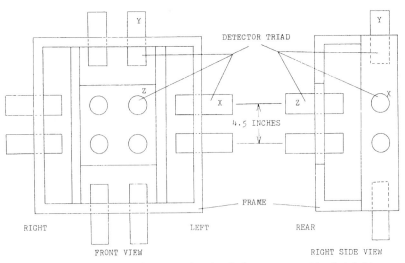

Figure 6.6 The detector configuration for the Orthoscanner or Vanderbilt University Mark 2 machine. The machine comprised four triads of three detecting elements, locked on to a rigid frame. (From Pickens (1977).)

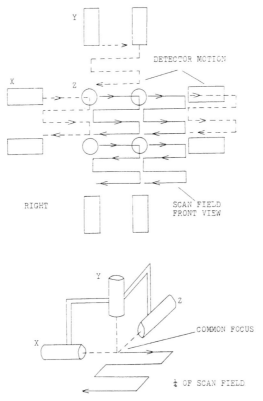

Figure 6.7 How the Orthoscanner achieved sensitive point tomographic scanning through a cube of side 9 in. (From Pickens (1977).)

The feasibility study for the machine was complete in 1971 (Pickens *et al* 1971) and the first patient was imaged on the completed machine (figure 6.8) in 1976. Pickens' (1977) *MS Thesis* contains a very thorough study of the engineering design principles of the machine. Linear movements were achieved via electric motors driving power screw threads (no back-hauling), ball bushings carrying the load. A major design constraint was for a small-size low-power unit. A DEC PDP9 provided control with data taking via CAMAC. During its life, several modifications were made including an upgrade from the original 37-hole collimators to lower-resolution 19-hole collimators with an eight fold improvement in sensitivity. Scanning movements in the z direction were not originally conceived and were added later in the project. These enabled direct sagittal and coronal scans. Several tomographic reconstruction methods were employed including a comparison of simply summing the output from detectors in each triad with using three times the minimum count in each triad. Results from the latter contained less 'tailing' artefacts. Pickens

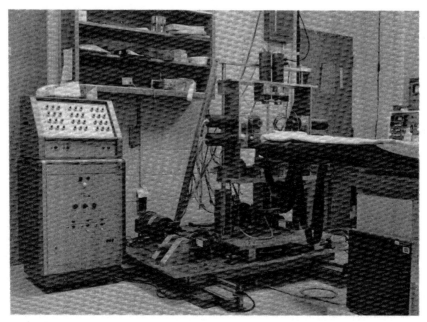

Figure 6.8 The Vanderbilt University Mark 2 Orthoscanner with stretcher and head holder in position. (From Pickens (1977).)

(1977) showed the results of several brain scans made with the Orthoscanner. These correlated well with the images from a gamma camera and also with X-ray transmission CT images.

This machine was never patented (Brill 1989) and the concept, if not the explicit engineering, was used for other developments including the single-slice multicrystal scanner originally marketed by Cleon–Union Carbide and later known as the Harvard imager and currently the Strichman imager (Stoddart and Stoddart 1979).

6.2.7 *Section scanning at Aberdeen*

Two types of section imaging were developed in Aberdeen in the early 1970s and first reported by Myers *et al* (1972) at an International Atomic Energy Agency symposium in Monte Carlo. One device was similar to the Anger multiplane scanner and is described in §6.3.1. The other was a development rather similar to the first two-detector transaxial section scanner built by Kuhl. Two crystal detectors were mounted some 40 cm apart and fitted with focusing collimators. The detector assemblies were arranged so that they could rotate about the patient to different orientations and could move along linear tracks at any particular location.

In the language of modern CT the device was a first-generation rotate–translate scanner. Data were captured in small steps of 4 mm along the track and transferred to punched paper tape which was subsequently fed to a PDP8I computer so that 64×64 element section images could be produced. In order to visualize other sections, the patient was moved longitudinally between scans through the plane of scanning. The width of each section was measured to be about 15 mm and the resolution in the plane of section was some 20 mm. This figure depended on the angular interval for the rotation which was initially chosen as 12° and later reduced to 6°. This also reduced the star artefacts generated by hot regions. In the Aberdeen implementation an allowance for the interference between transaxial slices was made by a point-by-point subtraction process. Further experimental details were provided by Bowley *et al* (1973). In subsequent years, several versions of the Aberdeen Section Scanner (ASS) were produced (Evans *et al* 1986).

6.2.8 *The Schmidlin four-view camera tomographic imager*

A novel way to use a gamma camera to reconstruct tomographic data was developed by Schmidlin (1972). The camera was fitted with a parallel-hole collimator and arranged with the camera face at 35.3° to the horizontal. This is the angle which the diagonal of a cube makes with the horizontal. Images were taken at four orientations of the object relative to the camera with the camera axis 'looking' down each cube diagonal in turn. This was done by arranging for the table supporting the object to rotate about the vertical. Schmidlin then used an iterative reconstruction technique to create a $32 \times 32 \times 32$ cube of reconstructed voxels to represent the object and reported good separation of structures.

6.2.9 *Gamma–gamma coincidence tomography*

Certain radionuclides emit two gammas of different energies in cascade. Examples are ^{75}Se, ^{111}In, ^{48}Cr, ^{43}K and the emission is nearly isotropic. Other nuclides such as ^{52}Fe emit a gamma and a positron and isotopes such as ^{125}I and ^{197}Hg decay by electron capture with subsequent emission of an X-ray and a gamma ray. This opens up new possibilities for tomography whereby coincidence detection may be employed (Powell *et al* 1970, Monahan and Powell 1974). The method is illustrated in figure 6.9. A pair of gamma cameras are arranged at right angles to each other. One is fitted with a standard parallel-hole collimator whilst the other has a slotted collimator. When a decay occurs, the former camera can determine the line along which the disintegration occurred and the latter gives the plane. The intersection of line and plane thus gives the point of emission. The limitations of such a method are those common to positron

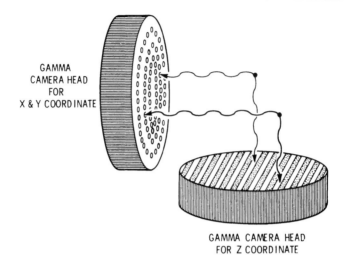

GAMMA
CAMERA HEAD
FOR
X & Y COORDINATE

GAMMA CAMERA HEAD
FOR Z COORDINATE

Figure 6.9 How two gamma cameras operating in coincidence—one fitted with a parallel-hole collimator and the other with a slotted collimator—can be used for tomographic imaging. The camera on the left gives coordinates (x, y) and the z coordinate comes from the camera on the right. (From Monahan and Powell (1974).)

tomography, namely the need to maximize the real-to-random-coincidence ratio and the need for large numbers of events. Since the coincidence ratio is inversely proportional to the source strength, the only way to achieve reasonable count rates is to reduce the resolving time of the detectors. In the equipment built by Monahan and Powell, a 10% real-to-random ratio was achieved with the resolving time of 30 ns. The (phantom) studies were very long (some 60 h) and it was suggested that the use of organic scintillators with shorter resolving times might allow increased activity without damaging the signal-to-noise ratio.

6.2.10 Multicrystal section imagers; from history to modern practice

We have seen that the beginnings of section imaging lay with arrangements by which one or more single-crystal detectors were moved around the body, from the earliest work by Kuhl through the development of the Aberdeen instruments. The Vanderbilt University machines utilized different principles (reinforcement of signals) but were also based on scanning detectors. The advent of gamma-camera-based tomography might have been expected to be the death of the other avenue of progress but this has in fact not been the case and there are a number of strong advocates even today of non-camera-based section imaging.

240

The reason lies in the greater sensitivity which can be achieved with scanner-based section imagers in view of the ability to place essentially a greater crystal area around the patient. The price paid is that generally these devices are single-section (or at best few-section) imagers and do not provide the full three-dimensional capability of camera-based SPECT systems. In a study of the origins of section imaging it is not appropriate to spend much time on the more modern instruments but some of these are listed in table 6.3 and they have been reviewed in some detail by Croft (1986) and by Ott *et al* (1988).

Table 6.3 Multicrystal section imagers.

Kuhl Mark 1–Mark 4
Vanderbilt University Mark 1 and Mark 2
Aberdeen Section Scanners ASS1 and ASS2
Tomogscanner (J and P Engineering)
Cleon 710
Harvard head scanner (improved Cleon 710)
Harvard body scanner (improved Cleon 711)
SPRINT (University of Michigan)
Tomomatic 64 (Medimatic)
HEADTOME1 and HEADTOME2
MUMPI
ASPECT

6.3 Emission laminography (focal-plane tomography; longitudinal-section tomography

As well as scanning-based and camera-based imaging of transaxial sections a number of important developments were going on at the same time in longitudinal tomography. These are the concern in this section.

6.3.1 *The Anger multiplane scanner*

Anger (1968) developed a multiplane scanner based on the principle that, when a three-dimensional distribution of activity was viewed with a scanning focused camera, the images of sources at different depths move at different speeds across the detection plane. For example with reference to figure 6.10 we may imagine there to be 25 point sources of activity at the 25 labelled locations, with five in each of five planes. These 25 sources would create the images shown. If now we imagine a point source moved

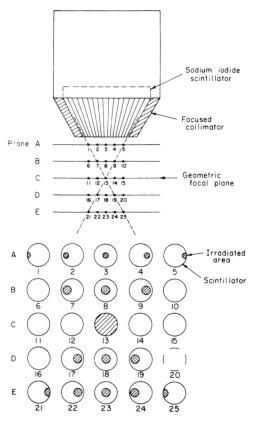

Figure 6.10 The method of operation of the Anger multiplane scanner explained in the text. (From Anger (1968) in *Fundamental Problems in Scanning* ed A Gottschalk and R N Beck. Courtesy of Charles C Thomas, Publisher, Springfield, Illinois.)

from point 1 to location 5 with the detector stationary, the image of this point would appear to move across the face of the detector. The motion which would be observed for a source moving from point 21 to 25 is similar except in the opposite direction. If a point moved from location 6 to 10, then its image would move faster across the detector and indeed would be outside of the field of view at locations 6 and 10. A source moving from 11 to 15 would only briefly be recorded (at position 13 when it would fill the detector). Hence the differential movement of sources at different depths is apparent. The same differential movement would arise if the source distribution were stationary and the detector executed a scanning motion.

Anger arranged for there to be five memory-core systems. For each scintillation at location x', y' in the camera when the camera scanning coordinates were x, y, an event was added to location $X_n Y_n$ in the nth image store where

$$X_n = x + knx'$$
$$Y_n = y + kny'$$

with $n = -2, -1, 0, 1, 2$ and k adjusted to select the separation between the planes. By this electronic reregistration of the data, exact compensation was made for the movement of images not in the focal plane such that in the data storage register corresponding to such a plane the image appeared not to move. In this manner, other planes were selectively focused. Only in the focal plane of the scanning camera was no reregistration needed ($n = 0$), this corresponding to the conventional scanning mode.

Anger originally proposed a somewhat complicated optical arrangement for achieving exactly the same effect. The principle is, however, clearer in the above form which became the basis of the Searle PhoCon imager (Anger 1969). Anger (1974a,b) reported further improvements to the multiplane imager as well as an investigation of other forms of motion tomography including a method by which the gamma camera executed a hypocycloidal motion not unlike that in the first work done by Ziedses des Plantes (figure 6.11).

PROPOSED TOMOGRAPHIC CAMERA

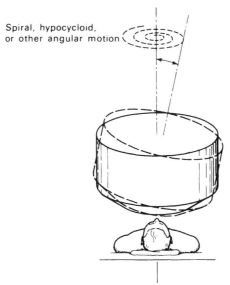

Spiral, hypocycloid, or other angular motion

Figure 6.11 A proposed hypocycloidal form of longitudinal-section tomography with a gamma camera, which is reminiscent of early X-ray focal-plane imaging. (From Anger (1974a).)

The very first Anger multiplane scanner was constructed at the Donner Laboratory using an Ohio-Nuclear frame and a model 8D focused collimator. The prototype is shown in figure 6.12. Anger (1974a,b) pointed out that the tomographic effect is achieved by *dispersal* of out-of-focus contributions to the image rather than *removal*. The residual cross-talk between planes was a problem (for all laminar tomography techniques) and characterized this class in contrast with what could be achieved with true section imagers (see §6.2).

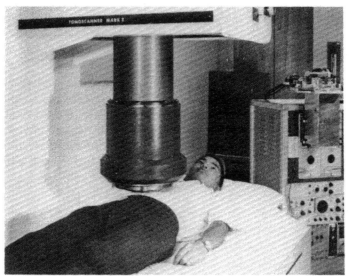

Figure 6.12 The prototype Anger multiplane scanner at the Donner Laboratory. (From Anger (1974b).)

A somewhat similar development was also made in Aberdeen by Myers *et al* (1972). In this implementation the detector remained stationary whilst the patient couch executed the required rectilinear scanning motion. Five planes were refocused by electronic techniques as described above in connection with the Anger device. The Aberdeen scanner was alternatively fitted with a collimator focusing to 18 cm and another focusing to 28 cm, this latter providing improved tomography in terms of out-of-plane blurring. The Aberdeen group chose to make this development rather than one based on a rotating collimator (see §6.3.3) because of the wider field of view which became available.

6.3.2 *The McAfee* et al *patent*

McAfee *et al* (1970–3) invented a new emission tomographic scanner which aimed to overcome some of the limitations of Anger's multiplane scanner using a focused collimator. The new method was first reported in 1969 (McAfee *et al* 1969) and a patent was granted in 1973.

The technique of McAfee *et al* may be understood with reference to figure 6.13. A gamma camera was provided with a slant hole collimator with all the holes at the same pitch. The camera–collimator combination made a linear scan recording the image dynamically, that is recording a sequence of images for each geometrical location of the camera relative to the patient. Thus the image of a radioactive region 9 appeared to move to

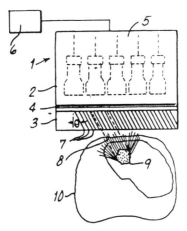

Figure 6.13 The arrangement used by McAfee *et al* for performing focal-plane laminography. A gamma camera 1 was fitted with a slant-hole collimator 3 which viewed a radioactive region 9 in a patient 10. The collimator orientation remained fixed as the camera translated with data being recorded dynamically. After one linear 'pass', further passes were made with the collimator at different orientations. Combining the data electronically from these scans led to selective slice focusing. (From McAfee *et al* (1970–3).)

the left in the camera frame as the camera translated right. After one complete lateral traverse the collimator was rotated 180° about the central axis of the camera so that the holes pointed in the opposite direction. A second linear traverse was then executed, data being captured in the same way. The advantage of this scanning method was that tomographic information could be digitally reconstructed after the scan. The means to do this is very reminiscent of early classical X-ray tomography except that the technique was an electronic analogy.

In figure 6.14 are drawn just two channels of the collimator in a position which they occupy on the first traverse A and B and on the second traverse G and F. Imagine that the data recorded have been converted back into fixed laboratory coordinates. The data from the two

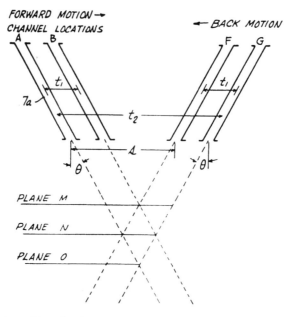

Figure 6.14 How the data from two scans with the slant-hole collimator at opposing directions could be combined for selective slice focusing. See text for explanation. (From McAfee *et al* (1970–3).)

scans can now be combined to accentuate the response to a particular plane. For example, if the data from A and F are combined and the data from B and G are combined, the radioactivity in plane N will be accentuated. Each channel alone has of course very little depth response since it can accept radiation from any depth along its field of view. However, by combining data, selective reinforcement of some particular plane can be arranged whilst the information from other planes is blurred. For example, in the case developed, channels A and F are not viewing the same location in plane M.

If, however, the combination is differently arranged, new planes can be brought into focus. For example adding channels A and G accentuates plane O whilst adding channels B and F accentuates plane M. McAfee *et al* arranged for the conversion of data into fixed laboratory coordinates and for the data combination to be performed electronically after the scan. In this way any number of slices at different depths could be reconstructed and displayed. The concept was very much like a digital version for nuclear medicine of Ziedses des Plantes' seriescopy (see § 1.3.2).

Combining data from more than two linear traverses increases the sensitivity of the method and in the patent a technique for achieving four traverses was described. These were orthogonal scans forwards and backwards (figure 6.15). Alternatively eight or 16 scans could be

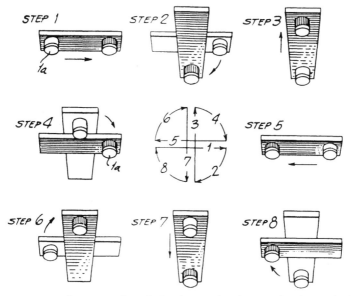

Figure 6.15 The manner by which selective slice focusing by the method of McAfee *et al* could be improved by combining data from four linear scans. (From McAfee *et al* (1970–3).)

performed along spokes of a wheel centred on the camera axis. The trade-off was between increased scan time and increased sensitivity (Stabler 1974).

Any type of gamma camera could be used, for example those patented by Tobias and Anger (1953–7), Anger (1958–61) or Blum (1971–5).

6.3.3 *Rotating-collimator tomography: Freedman, Walker and Muehllehner*

We have seen that, although the work of Kuhl and colleagues in 1963 included developing methods for single-section transverse axial tomography, many of the other developments in the 1960s were still aimed towards focal-plane or longitudinal-section tomography. In this respect they attempted to enhance the effect of the signal from some specific plane whilst signals from other planes were blurred out. The earliest notable developments were the Anger multiplane scanner and the McAfee slant-hole collimated scanner. These inventions were analogous to classical tomographic techniques in transmission X-ray imaging as reviewed in chapter 1 and therefore aimed only at improving detectability by blurring unwanted signals. Perhaps ingraciously it might be remarked that they suffered the same kinds of artefactual defect.

Several workers invented longitudinal-section imagers which departed from the concept of a *scanning* detector, instead centring their techniques on the use of a camera fitted with a *rotating* collimator. In the simplest of such inventions, the camera itself did not move (see §6.3.3.1) whilst, in more advanced implementations, compound motions of both collimator and detector were developed (see §6.3.3.2). This was done in order to improve the sensitivity of the laminar-focusing method, in the case of Muehllehner's equipment by a factor of 3 over the Anger tomographic scanner. A reduction in resolution was, however, a consequence. The compound motion equipment was able to achieve a field of view at least as large as the camera area whilst, if only the collimator rotated, the field of view was a truncated cone with base equal to the camera area and vertex at a distance equal to the camera radius times the cotangent of the collimator-hole inclination. The key developments in rotating-collimator tomography will now be separately considered.

6.3.3.1 The Freedman tomographic camera and patents by Walker and by Muehllehner. The invention of the tomographic camera employing a rotating collimator was disclosed in a patent in 1968 by Walker (1968–71). Similar proposals were embodied in a paper by Follett in 1967 (Trott 1969). Freedman (1970, 1972, 1974) developed the idea. He attached a collimator with parallel angled holes, inclined at $\theta = 20°$ to the vertical (manufactured by the Atomic Development Corporation) to a Nuclear Chicago Pho/Gamma 3 camera. The collimator did not rotate continuously during data acquisition; instead it stopped at 12 finite orientations 30° apart. The principle of rotating-collimator tomography is very simple. Any particular source point of emitted single photons is registered as an event point at the camera which executes a circular locus as the collimator rotates. In Freedman's implementation, instead of a continuous circle of detection points, 12 finite event points on a circle would have been registered. The radius of the circle is simply the depth of the source point multiplied by the tangent of the angle of inclination of the collimator holes. In order to focus a particular plane selectively (at the expense of defocusing the others), the data are electronically reregistered. This is achieved by rearranging the separate camera views such that data within each view are moved to new locations, and the separate rearranged data superposed. The algorithm for reregistration is simply a function of the collimator angle, the orientation angle of the collimator and the depth of the plane which is desired in focus. Essentially data are combined by grouping points which lie on the circular locus previously described and, by varying the radius of this grouping, different planes may be brought into focus. Readers wishing to see the equations involved might consult the patent by Walker (1968–71) or Muehllehner (1973–4). Alternatively the paper by Freedman (1974) has a diagrammatic explanation. Freedman transferred the camera data to magnetic tape for reconstruction by computer. In the Walker patent the tomographic images can be created directly by introducing electronic offsets into the position signals whereby one selected plane may be brought into focus at the expense of others. Alternatively, by providing a plurality of such circuits each adjusted for different offsets, whole sets of tomographic data may be generated simultaneously. The Walker patent described continuous rotation of the collimator rather than the multiple-fixed-orientation method experimentally developed by Freedman.

It is obvious that, when a particular plane is reconstructed by this technique, data in other planes fail to coordinate properly and are consequently blurred as required. A point a distance g from the plane in focus will be imaged to a circle of radius g times the tangent of the collimator slant angle. This strong defocusing is the foundation of the method. However, similar criticisms of such circular blurring were levelled by Freedman to those levelled at circular motion for X-ray tomography. Highly localized hot regions can create artefacts such as haloes. Freedman remarks that, if movements such as elliptical or hypocycloidal motion could be engineered, the blurring would be more complete and less prone to artefacts. We recall similar comments made for the various movements in classical X-ray tomography.

The principal disadvantage of Freedman's method is the decreased field of view and hence lack of uniform sensitivity with depth. The field of view seen at all angles of the collimator is a truncated cone. This was a big limitation at the time because of the smaller size of gamma cameras compared with today. We shall see in the next section what was done to overcome this. Several workers attempted to improve the spatial resolution of the tomographic technique by rotating a magnifying collimator (see, for example, Rudin *et al* 1974). Such a device sacrifices field of view and is essentially a hybrid of the rotating parallel-hole system and the non-tomographic magnifying collimator.

6.3.3.2 The Muehllehner compound tomographic images. Muehllehner invented compound-motion laminar tomography. His first equipment was similar to that of Walker (1968–71) and Freedman (1970), namely a slant-hole collimator with all holes parallel attached to a gamma camera such that the collimator could rotate (Muehllehner 1971, 1974). The new development was to arrange for the table supporting the patient to precess slowly (approximately 1 rev min^{-1}) and in complete synchronism with the rotation of the collimator. The table moved in a circular motion about the central axis of the camera–collimator at a fixed radius of rotation and in such a way that it did not rotate about itself. This motion can be termed circular precession. In this arrangement there is a single plane, which was termed the 'mechanical focal plane' in which each position is viewed throughout the data acquisition by some particular

collimator hole. The image of each point in this plane remains stationary while the collimator rotated—hence the name given to this plane.
A stationary image of the mechanical focal plane was constructed with no
resort to reregistering signals. The distance to the focal plane was simply
given in terms of the collimator angle and the radius of precession of the
table. The advantage of this over the Freedman equipment was that the
useful area of uniform sensitivity was equal in this plane to the useful area
of the camera. The images of points in all other planes executed circular
loci in the plane of detection. However, by reregistering the data in a
circular fashion in synchronism with the rotation of the collimator, other
planes could be selectively focused. By varying the radius of the circle of
reregistration different planes were brought into focus. This method was
patented (Muehllehner 1970—2).

Later, Muehllehner (1973—4) patented a different method (at which he
had already hinted in the 1971 paper) which relied on similar principles.
In this, instead of the patient executing circular precession, the camera was
given this movement and the patient remained stationary. Once again the
angled-hole collimator rotated in synchrony with the precession of the
camera. In figure 6.16 the theory is clearly shown. In this arrangement
there is also a naturally focused plane Db in the figure which is the plane
defined as containing the point B at which the axis of precession of the
camera intersects the line from the centre of the camera passing through

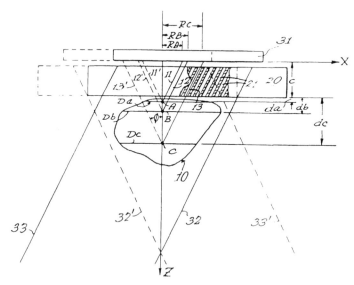

Figure 6.16 A side view through a system for rotating-collimator
tomography with camera precession. This complex figure is explained in
the text. (From Muehllehner (1973—4).)

one of the angled holes. The camera is shown at 31, the collimator at 20, an object being imaged at 10, the z axis is the axis of precession of the camera (which of course does *not* rotate about itself), RA is the radius of this precession and is the distance from the axis z to the centre of the camera. The collimator holes are inclined to the vertical at angle ϕ. The full lines represent one particular orientation of the camera and collimator, and the broken lines are the positions occupied half a rotation later. It is thus apparent how plane Db remains focused. Points in other planes, such as A and C, are imaged as circular trajectories. However, these planes can be selectively focused when the data are reregistered by circular precession in the way described earlier. At the time of the patent this was engineered via analogue circuits rather than digital circuits. The equations for reregistering are

$$X = x + (Z \tan \phi - R) \cos \theta$$

$$Y = y + (Z \tan \phi - R) \sin \theta$$

where X and Y refer to the stationary coordinate system into which data are combined (that corresponding to the axis Z through the centre of precession), x and y are the local axes on the camera (which precess at radius R about axis Z), θ is the angle of precession and ϕ is the collimator slant angle. From these equations the lack of need to reregister data when $R = Z \tan \phi$ is apparent (that is for the naturally focused plane). To focus other planes selectively, the radius for reregistration is altered for different distances Z. In passing, we might also note that the Walker and Freedman camera can be thought of as a special case where there is no camera precession ($R = 0$). In this case there is no naturally focused plane and to bring any plane into focus requires reregistering data via the reduced form of the above equations with $R = 0$. Also we see how the radius of this precession in reregistering data has to increase for increasing depth of plane. The dependence on the tilt angle ϕ of the collimator is also clear. This determines the 'strength' of the tomographic effect.

Once again the advantage of this arrangement was increased field of view with constant sensitivity. It achieves this by combining one of the features of the Anger device (movement of the camera) with that of the Walker device (collimator rotation). Both of Muehllehner's implementations of compound-motion laminography do, however, require quite complex engineering of equipment to provide the circular trajectory. It was to avoid this that Myers *et al* (1972) at Aberdeen chose to build instead a scanning detector (like Anger's) with rectilinear movements (see §6.3.1). Gamma cameras with rotating collimators were often called 'tomocameras' or systems for ring tomography in that out-of-focus information is blurred into rings (Anger 1974b). This dispersal mechanism is the physical basis of the tomographic effect. The Muehllehner collimator became available commercially as an attachment to the Nuclear Chicago camera. Systems with rotating collimators, however, did not find wide clinical acceptance (Keyes 1982) although they received renewed interest later on when reconstruction methods became better developed (Chang *et al* 1980).

Muehllehner's (1988) reflective comments run as follows: 'Shortly after Anger's development, a number of people described similar ideas using a stationary camera, and, in June of 1970, both I and Dr Freedman in completely separate efforts described a stationary camera with a rotating collimator coupled to a moving bed in order to do longitudinal tomography. Our two papers were so similar that we could have easily exchanged our slides; yet we had never spoken to each other prior to this meeting. Longitudinal tomography with rotating collimators was not well received by the clinical community, since it did not significantly improve the diagnostic accuracy of nuclear medicine procedures, but was more complicated than planar imaging.'

A system similar to that invented by Muehllehner was designed and built by Cottrall and Flioni-Vyza (1972). In their embodiment of the principles, they had the patient couch execute the planetary motion with the camera stationary. The camera was fitted with a 45° slant-hole rotating collimator to increase the tomographic effect.

6.3.4 *The Miraldi Tomoscanner*

In 1969, Miraldi began work on a new form of tomographic scanner which later became known as the 'Tomoscanner', a name unfortunately also used for other devices such as that of Brill *et al*. The scanner was announced in New Orleans (Miraldi *et al* 1969). The development of the scanner was described over several years in papers (Miraldi and Di Chiro 1970, Miraldi *et al* 1974) and the technique was patented (Miraldi 1971–4). The patent possibly provides the clearest description of the physical basis of the method.

The Miraldi Tomoscanner comprised a bar of scintillation crystal viewed at both ends by a photomultiplier tube. The detector was provided with a collimator which was unfocused along the length of the strip detector but focused in the orthogonal direction as shown in figure 6.17. Essentially the detector viewed a wedge-shaped region of space. The logarithmic amplitude of the output of each photomultiplier was found to be precisely linear with position of the scintillation along the length of the detector and this was used to give the strip detector its position sensitivity. The whole assembly was translated at constant speed across the object being imaged. The data from this single scan of course possessed no tomographic information other than the very weak effect of all naturally focusing collimators. The tomographic effect was achieved by

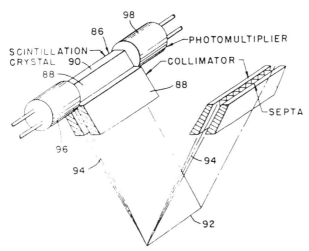

Figure 6.17 The principle of operation of the Miraldi Tomoscanner. The detector is shown in two of its orientations. The scans were made in the plane of the paper and then at sequential orientations in the plane normal to this. (From Miraldi (1971–4).)

reversing the tilt of the collimator symmetrically about the normal to the scanning direction as shown by the other position of the assembly in the figure, and repeating the scan in the reverse direction. By digitally summing the two scans, the plane defined by the intersection of the vertices of the two wedges was reinforced whilst the response from emissions in other planes was blurred out. The tomographic effect was increased by repeating this doublet procedure with the whole assembly rotated in the plane of scanning in angular increments around a full circle. In general, 24 such double scans were made at 15° increments. After correcting for the rotation the data were superposed. In the first experimental scanner the holes of the collimator were inclined at 45° to the vertical.

Miraldi's principle could also be achieved of course by scanning a pair of collimated assemblies, with the collimators pointing the two different ways in each and the assemblies being offset by some fixed distance. Figure 6.17 could be viewed in this way if preferred. In this case the repeated scans need only be made over a half-circle. The tomographic effect is achieved in the same way. After scanning, other planes could be selectively focused by combining the data in different ways. It is easy to see that, if the two assemblies were closer, the plane in focus would be nearer the plane of motion of the detectors and vice versa. Hence by moving one data set with respect to the other (scanning having been done with the detectors at a fixed separation), different planes may be brought into focus. This was achieved in the Miraldi Tomoscanner by reregistering the data offline in a computer after the scans had been made. The principle of reregistering images to focus different planes selectively is one common to most of the longitudinal tomographic imagers described in this chapter. The difference between the imagers is more concerned with the manner of acquiring the individual data sets (scanning single detectors versus scanning cameras versus fixed cameras with rotating collimators versus compound-motion cameras, etc). The reinforcement of data in the selected plane gives the tomographic effect at the expense of blurring detail from the other planes. This flexible *a posteriori* superposition of nuclear data is closely allied to the seriescopy technique in classical X-ray tomography invented by Ziedses des Plantes. Of course the X-ray seriescopy developments were purely analogue and more was possible with the use of digital reregistration.

The choice of collimator for the Miraldi Tomoscanner was made after a number of alternative and quite different types were considered (and described in the patent). One was a circular collimator comprising annular rings of holes focused at different planes in the patient. This was rejected because it would have required separate scintillating rings or a single ring with Anger-like logic to select signals from given radii in the detector. Another was a strip detector in which holes converged to different planes in the orthogonal direction and were parallel along the strip. This was also rejected because it would also require a hybrid detection scheme.

The manufactured Miraldi Tomoscanner (Miraldi *et al* 1974) comprised a single detector mounted on a C arm above a couch. The detector assembly could be rocked to vary the inclination to the vertical of the collimator and the whole assembly was rotatable azimuthally in the plane of scan. In practice therefore (not having a dual detector) a full 360° rotation was performed. In later clinical work (Di Chiro *et al* 1974) it was found advantageous to reduce the collimator tilt from 45° to 30°. Of course by setting zero tilt the scanner could function as a conventional non-tomographic imager. Because of the straight-line relationship between the logged output from each single photomultiplier tube with distance of the scintillation from the tube (which was the basis of the detector's positional sensitivity), the sum of the outputs was essentially constant. Thus very precise pulse height discrimination was possible in order to minimize the effect of Compton scatter interactions in the detector.

6.3.5 Coded aperture imaging

Before leaving the subject of longitudinal focal-plane tomography it is appropriate to mention the attempts that were made in the early 1970s to develop techniques based on the use of a coding aperture. Perhaps the most commonly known method was the attachment of an X-ray Fresnel zoneplate aperture to a gamma camera in the place 'normally' occupied by the collimator. The aperture casts a shadow of itself onto the detector generating a coded image of the source distribution. The coding is three dimensional because the size of the shadow depends on the distance of the source from the detector and the positional coordinates (x, y) on the detector depend on the corresponding coordinates in planes of the object parallel to the detector face. Other apertures which were investigated were those comprising random or regular patterns of pinholes as well as time-varying patterns of transparent and opaque segments. In all cases the principle was the same; that doing away with the collimator could greatly improve efficiency. Not only could an Anger gamma camera be employed but also image intensifiers, multiwire proportional chambers and film (Barrett *et al* 1974). The difficulty with these systems included the need for rather complex decoding procedures which were executed either in analogue fashion using coherent light or on a digital computer. Unfortunately the depth resolution was fairly poor especially with the Anger camera and there was a tendency for information from adjacent planes to interfere with the plane in focus. Although the techniques developed had a certain elegance, they never truly rivalled other single-photon techniques and were rather overtaken by the development of SPECT.

6.3.6 Seven-pinhole tomography

Entering late onto the scene of limited-angle tomography the use of a collimator with seven pinholes has been described as an idea which arrived in the right place at the right time (Keyes 1982). The collimator provided seven views of the distribution of activity from separate directions and these could be used to reconstruct a set of planes parallel to the collimator via a small digital computer. The success of the method (Vogel *et al* 1978) was due to the availability of wide-field-of-view gamma cameras coupled with reconstruction experience from the explosion of interest in CT and the widespread use of the digital computer. The development was well matched to the need for ^{201}Tl myocardial imaging, so much so that this is one of the few techniques for longitudinal emission tomography which is still widely in use despite the availability of SPECT. The method is only suitable for imaging small organs such as the heart.

6.4 Positron tomography

6.4.1 Early medical experiments

So far this chapter has concentrated on the origins of imaging techniques which display the spatial distribution of radionuclides *in vivo* which decay with single-gamma emission (touching briefly also on cascade imaging).

Because each single photon carries no specific spatial information, it is only possible to create three-dimensional tomographic maps by making use of a collimator with some form of movement between the detecting device and the patient.

In this section the origins of PET are highlighted. The method is inherently different. Certain radionuclides decay with the emission of positive electrons (positrons). Within a short distance (approximately 1 mm) of the point of emission the positron annihilates with an electron and yields two 511 keV single photons which are emitted approximately back to back. If these can be accurately detected it is apparent that the original decay which led to their production must be somewhere along the line joining the points of detection. This principle is the basis of PET.

In 1951, Wrenn *et al* (1951) published the first study of positron counting with medical possibilities. They arranged for two scintillation detectors (1 in crystals of sodium iodide) without a collimator to scan across a distribution of the positron emitter ^{64}Cu placed within a fixed brain contained in its skull. The detectors were arranged opposite to each other and passed either side of the brain. The outputs of the detectors were in coincidence, that is to say an event was recorded only when both detectors simultaneously registered a count. They showed that the width of the profile corresponding to a point emitter was significantly narrower than the corresponding measurement made by just one of the detectors counting single emissions. The background to their work was the desire to create a method for externally detecting tumours in the brain by recording the differential uptake of a radiopharmaceutical and possibly to replace the single-photon technique based on ^{131}I which was the most popular at the time. The early work did not create three-dimensional images but it laid the foundations of PET.

At much the same time work at Massachusetts General Hospital (MGH) (Sweet 1951, Brownell and Sweet 1953) was under way with similar aims. Two sodium iodide detectors were coupled together in coincidence and arranged to scan in a raster pattern. Planar scans of the head were made. This group went a little further and arranged for the signals from each detector to pass into a circuit which detected the difference in the average counting rates of the two detectors. In this way a crude measure was made of whether the source of emission was nearer one detector than the other. The coincidence scans were made at the same time as the so-called 'unbalance scans'. The sign of the difference signal was used to determine which symbol was used on the attached printer for the radioactive distribution (Aronow 1966) (see figure A.4 in the appendix). Again this was not true tomography but it was a move in the right direction. Figure 6.18 shows the first positron scanner in 1952 in the laboratory. Figure 6.19 shows the first clinical positron scanner (1955) with Gordon Brownell standing and Saul Aronow leaning over the scanner and figure 6.20 is an

Figure 6.18 The first double-crystal positron scanner produced by Brownell photographed in 1952. (Photograph courtesy of G Brownell.)

image taken with this system. Figure 6.21 shows the commercial version of the double-headed scanner in about 1956. In the text *Instrumentation in Nuclear Medicine* edited by Hine, which has since become regarded as a classic description of the instrumentation of the mid 1960s, Brownell *et al* (1966) wrote, 'At the present time no practical technique for true three-dimensional visualisation of radioisotope sources has been developed.'

According to Sweet (1951) the work of Wrenn *et al* at Duke University had started a little earlier than that at MGH. Sweet was a neurosurgeon at MGH, and Brownell a physicist from Massachusetts Institute of Technology and hospital physicist at MGH. Prior to developing brain scanning with positron emitters, a good deal of success at locating tumours intraoperatively was obtained with a very small Geiger–Müller counter recording locally the beta emission from ^{32}P which preferentially locates in brain neoplasms and is not much taken up by normal brain tissue. Some experience was also gained in externally counting single gammas from ^{42}K but with less success. Sweet's words that the procedure was like the use of a colorimetric technique for finding a black cat behind a pile of coal summed up the problem. Conceptualizing positron imaging came up during a trip to Brookhaven by Dr Sweet and Dr Brownell in 1950 and shortly thereafter the positron imaging system described by

Figure 6.19 The first clinical positron scanner (1955) from MGH. Gordon Brownell is standing with Saul Aronow adjusting the machine. (Photograph courtesy of G Brownell.)

Sweet was constructed. Brownell (1988, 1989) believes this was the first use of positrons for medical imaging and reaffirmed the independence of this work from that of Wrenn, Good and Handler.

Before moving on to describe the earliest instrumentation for positron tomography, it is important to understand the driving forces behind the urge to image high-energy radiation. These centred on the synergistic combination of a number of factors. First was the new found availability of the 'physiological radionuclides' ^{11}C, ^{13}N, ^{15}O and ^{18}F which were produced by cyclotron irradiation of stable elements. During the late 1930s and through the 1940s a number of metabolic studies were carried out using these short-lived tracers. However, their short half-lives were considered to be a disadvantage and, by the early 1950s, biomedical studies had dwindled. It was not until the middle to late 1950s, when it was realized that they offered a particularly attractive method for the regional study of metabolism, in view of the presence of these elements in biomolecules, that interest revived and early assessments that, for

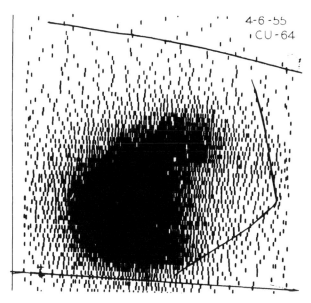

4-6-55
CU-64

Figure 6.20 Image of a dog liver taken in 1955 with the double-crystal positron scanner and the radionuclide ^{64}Cu. (Photograph courtesy of G Brownell.)

Figure 6.21 The commercial version of the first clinical positron scanner (circa 1956). (Photograph courtesy of G Brownell.)

259

example, ^{15}O was unsuitable, were found to be unduly pessimistic (Ter Pogossian 1985, 1988). This realization, combined with greater availability, was exploited by imaginative chemists to label complex molecules with physiological labels rapidly despite their short half-lives. Indeed the short half-lives became an advantage since relatively large quantities of radioactive material could be administered without dose penalty. About the same time, rapid developments in instrumentation took place using the advantageous properties of the positron annihilation. Making use of the 'time of flight' of the back-to-back gammas was, however, not possible until quite late in the story of positron tomography when Ter Pogossian and colleagues were able to develop the generation of machines known as PETT and Super PETT (Ter Pogossian *et al* 1982, Ter Pogossian 1985). The PETT series of PET scanners had their origin with modifications of the Kuhl Mark 4 scanner (see §4.3) for coincidence detection. This arose from the cooperation which stemmed from a review by Ter Pogossian of Kuhl's grant applications for the Mark 4 scanner and its application to measuring cerebral blood volume, flow and glucose utilization in 1973 (Kuhl 1989). Before these modern machines, a number of imaginative imagers were constructed. With regard to the use of oxygen as a physiological tracer the remarks of W E Siri (1949) and M D Kamen (1957) have been much quoted. These were (respectively) as follows: 'The unstable species of longest half-life is ^{15}O (126 s); this has not been employed for tracer work and does not offer much promise.' and 'No radioactive isotope of oxygen is sufficiently long lived to be useful in tracer work.' Ter Pogossian and Herscovitch (1985) wrote, 'It is ironic that today ^{15}O is a radioactive tracer which has proven itself to be of major importance for the *in vivo* regional measurement of blood flow, blood volume and oxygen metabolism in the brain and other organs.' The use of ^{15}O was pioneered at Washington University in St Louis and at London's Hammersmith Hospital as well as at other centres noted later.

6.4.2 *The MGH positron camera (Brownell* et al)

Positron counting continued to be developed for many years at MGH through generations of counters and scanning detector systems. The first true positron *camera* at MGH was not completed until the late 1960s (Brownell *et al* 1969, 1972, Brownell and Burnham 1974b). This comprised two planar detectors which were arranged either side of the patient (figures 6.22 and 6.23). Each detector comprised 127 crystals of sodium iodide of circular cross-section (2 cm diameter) and 3.8 cm deep which were viewed by eight rows and nine columns of photomultiplier tubes. The scintillators and tubes were so arranged that each tube received light from four crystals and each crystal was divided between two tubes (figure 6.24). This coding of the data led to considerable saving in the cost

Figure 6.22 The prototype positron camera at MGH. (From Brownell *et al* (1972).)

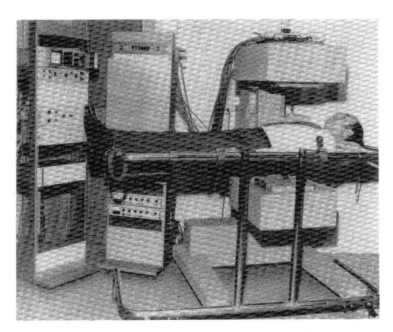

Figure 6.23 The first area-sensitive positron camera at MGH circa 1968. (Photograph courtesy of G Brownell.)

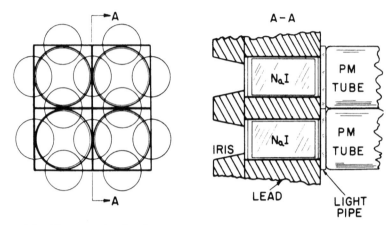

Figure 6.24 The principle whereby crystals could share photomultiplier tubes to reduce cost and the difficulty of close packing. The dark circles on the left are photomultiplier tubes and the smaller circles are the crystals. (From Brownell *et al* (1972).)

of tubes and is a principle much copied thereafter. Coincidence circuitry was arranged between the detectors such that each scintillator in one detector bank was in coincidence with 25 in the opposite bank. That is to say the coincidences were not restricted to the 127 paraxial pairs. Coincidence events were back-projected into five equally spaced planes between the detector banks and thus a form of tomography was possible. The camera was used for a very large number of physiological studies and complemented the programme at the hospital in the development of cyclotron-produced radionuclides. Figure 6.25 shows the first tomographic images of a dog heart.

If the coincidence channels had been restricted to the 127 paraxial set, no tomography would of course have been possible and the instrument would have functioned as a conventional camera. The tomographic effect is entirely due to the incorporation of non-paraxial coincidence channels. The restriction to the use of 25 scintillators coupled to each opposing scintillator led to diamond-shaped artefacts in the out-of-focus planes. Essentially, if a point was exactly located in one of the reconstructed planes, lines through that point and the single scintillator on one side to the 25 scintillators on the other side develop a raster of points lying in a diamond on the out-of-focus planes. In order to overcome this, Brownell and Burnham (1974a) introduced small motions into the detectors. They investigated the use of square motions, line motions and an octagonal motion approximating to a circle. The camera achieved a spatial resolution of some 6–9 mm and was used principally to image ^{11}C, ^{13}N and ^{15}O as well as ^{18}F. It would appear that in 1971 the camera was only just beginning to be used in tomographic mode. Brownell (1988, 1989) has

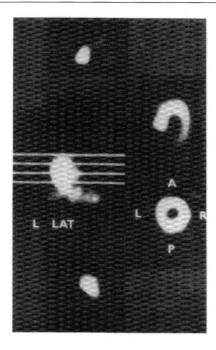

Figure 6.25 The first tomographic images from the MGH area positron camera (circa 1970). (Photograph courtesy of G Brownell.)

reported that, since the earliest days of positron counting, some form of positron imaging device has been operating continuously at MGH for either clinical purposes or research. The development of the positron camera built on the experience from the hybrid positron scanner with a pair of nine-crystal linear detectors and achieved a tenfold increase in sensitivity (Brownell *et al* 1969). The PC1 camera was not patented because it had been disclosed in several papers but it was covered by a *US Atomic Energy Commission Record of Invention* (S-40, 757) in November 1972 (Brownell 1988, 1989). This recorded the date of inception as June 1968, completion of the model in March 1969 and first tests in May 1971 although, as Brownell has pointed out, 'This evolved from a long series of positron imaging devices dating back to the early 1950s.' Patent protection for the next model (PC2) was also dropped for similar reasons.

6.4.3 *Positron imaging at the Brookhaven National Laboratory*

A different approach to positron imaging was taken at the Brookhaven National Laboratory in New York from that at nearby MGH. The story of the development shows the interesting link between positron scanning and section imaging. At first the obvious extension from double-crystal scanning to multiple-crystal scanning was conceived. It was then realized (Robertson and Niell 1962) that there was no reason to confine the scanning detectors to a plane and an arrangement whereby 32 detectors were arranged on the surface of a sphere was tried. There was then no need for movement in order to generate a three-dimensional distribution. The device was built and christened the 'non-inertial positron scanner' although it was known locally by some as the 'hair drier' (figure 6.26)! The machine was built by Higinbotham and Rankowitz in 1960 (Yamamoto 1988). The approach was abandoned when it became clear that 32 crystals in this arrangement could not give the required resolution and that data from a 32-crystal system, let alone one with more detectors, was too complicated to analyse. It was therefore decided to work with the 32 crystals arranged in a single plane. The system was designed for brain investigations and is shown in figure 6.27. The ring of 32 crystals was mounted on a circle 40 cm in diameter. The crystals were sodium iodide 3.2 cm in diameter. Each crystal detector was electronically coupled for potential coincidence with each of its 31 neighbours. For example with reference to figure 6.28, if a point source of emission was located at the darkened spot, the arrowed lines indicate the possible back-to-back photon paths which would lead to coincidence detection. The patient was

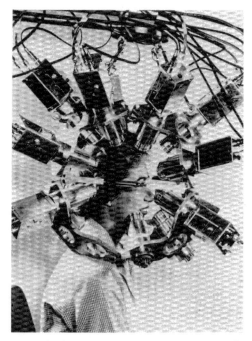

Figure 6.26 The 'hair drier'. (From Robertson and Niell (1962).)

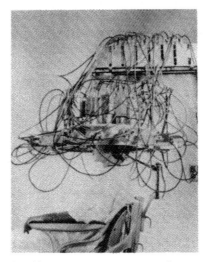

Figure 6.27 The Brookhaven prototype 32-crystal ring system positron camera—the 'head shrinker'. (From Robertson *et al* (1974).)

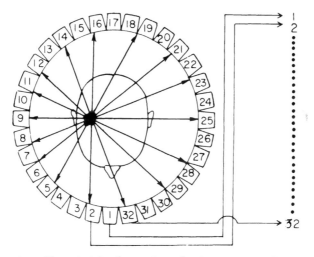

Figure 6.28 The principle of operation of a ring-system positron camera. (From Robertson *et al* (1974).)

seated on a dental chair for positioning (Rankowitz *et al* 1962, Robertson *et al* 1974). This machine became affectionately known as the 'head shrinker'. It was constructed in 1961 and very quickly replaced the role of the 'hair drier'.

From the coincidence data recorded, a planar section through the head can in principle be reconstructed. Essentially the coincidence counts in some particular channel (that is a pair of detectors) must relate to the sum of activity along the line joining the detectors with appropriate weighting for the geometric efficiency. The equipment development began early in the 1960s (Robertson and Niell 1962) but there were problems with the reconstruction. The first methods considered were matrix inversion techniques but these were not found to be very satisfactory (Bozzo *et al* 1968). Certain simplifications were then adopted. It is inherently obvious that, if coincidence channels between detectors with fixed separations around the circumference of the circle were considered, then these channels are insensitive to activity within a circle defined by the radius of a chord connecting any particular pair. For detectors with small circumferential separation the excluded circle has a large radius and, as the circumferential separation is increased, the circle of exclusion decreases in radius until it is just the central point for detectors at opposite ends of a diameter. This led to the development of an 'onion-peeling' algorithm for reconstruction (Robertson and Bozzo 1964, Robertson *et al* 1974).

Clinical results from the 32-detector 'ring system' were compared (for the same patients) with images produced by the two-detector MGH hospital and the results in the brain were shown to assist with diagnosis (Bozzo *et al* 1968). Brownell (1988, 1989) wrote, 'I must say that we never took the Brookhaven device very seriously. Its original aim was not tomography. . . . The instrument knocked around for quite some time and we had it here for awhile. I suppose, however, it could be considered as one of the original single plane ring devices and, of course, certainly pre-dated PETT and ECAT devices by many years.' The workers who developed the first ring system very early recognized that improved spatial resolution was possible by increasing the number and decreasing the size of the detectors and this has of course occurred in time until today huge rings of crystals are engineered (see, for example, Burnham *et al* 1982–5).

The 'head shrinker' which replaced the 'hair drier', the 32-crystal ring system constructed at the Brookhaven National Laboratory, was conceived in 1961. It enjoyed, however, two separate phases of development parted by a considerable hibernation of at least 5 years. Events have been explained by Yamamoto (1988). When it was first constructed, the interest lay in investigating brain tumours on the basis of disruption of the blood–brain barrier. However, the widespread use of the gamma camera and the introduction of $^{99}\text{Tc}^{m}$—labelled radiopharmaceuticals soon overtook these aims and the PET device was placed in storage at the Medical Department (Feindel and Yamamoto 1978).

In 1966, Yamamoto and Robertson proposed a new use for the 'head shrinker' to the Clinical Investigations and use of Radioisotopes Committee at Brookhaven that it form the imaging component for new studies of regional cerebral blood flow using the positron emitter ^{79}Kr. The technique was expected to be uniformly sensitive to all regions of the brain and associated with a low radiation dose. They proposed to investigate a range of the normal population and study the effect on regional cerebral blood flow of age, position of patient, sleeping versus waking and sensory stimulation. For pathological studies they proposed investigating patients with hypertension, Parkinson's disease, cerebrovascular disease, tumours and under medication. Their application was accepted on 14 July 1966 but no clinical work with patients was performed during the first 5 years, necessitating a reapplication by Robertson on 10 December 1971. The reason lay in the significant upgrades to the equipment made under Robertson's direction including redesigning the detector system and data capture. During this period, Yamamoto had moved to the Montreal Neurological Institute and Hospital (McGill University) and the request for reauthorization indicated that patients would derive from across the border, a fact commented upon by the reviewing committee. At this time, Robertson was Director of the Medical Physics Division at Brookhaven National Laboratory and correspondence shows that Yamamoto paid frequent visits to Robertson at the Laboratory to work on the improved equipment.

Yamamoto was keen to have the 'head shrinker' in Montreal and set in motion a train of events which led to this happening. On 2 October 1974 the Chairman of the Medical Department at Brookhaven, E P Cronkite, wrote to the Director of the Montreal Neurological Institute, W Feindel, agreeing to the transfer of equipment for a finite term. On 28 March 1975 the Associate Chairman, R B Aronson, authorized the transfer under a loan contract. Yamamoto drove to Brookhaven to effect the transfer on 6–8 April 1975. Work progressed well in Montreal, particularly as a collaboration formed with Yaffe to make use of other positron emitters including ^{77}Kr, ^{123}Xe, ^{124}I, ^{11}C and ^{68}Ga. The equipment was modified to operate vertically, to rotate by 6° to simulate data capture by 64 crystals, and to use different reconstruction techniques from those of Robertson and Marr. Antiscatter shielding was also added. The work supported nine staff under three funding agencies.

On 21 October 1975 a letter was sent from Brookhaven by H Atkins, requesting return of the 'head shrinker' but, after some correspondence between Yamamoto, Cronkite and Aronson, an extension to the loan period was agreed. Cronkite wrote that Federal Regulations applying to property of the US Government required annual renewal of loan agreements. Work flourished in Montreal (Yamamoto *et al* 1976, 1977) where the First International Symposium on Positron Emission Computed Tomography took place on 2–3 June 1978 (JCAT 1978). It is amusing to note this development of positron imaging took place in the very university where Sir Ernest Rutherford was Professor for nine years and who in 1914 thought he had discovered the positive electron.† By the mid-1970s this scanner became known as Positome-1. The 1 distinguishes it from a later development (Thompson *et al* 1978) Positome-2 which had 64 crystals of a different scintillator bismuth germanate (BGO). This machine was also put into service measuring regional cerebral blood flow with ^{77}Kr.

6.4.4 *Positron imaging based on Anger cameras*

Not long after the invention of the Anger camera, it was being used for positron imaging. A number of different geometries combining an Anger camera with one or more focused scintillation detectors were developed for early positron tomography. These were reviewed by Anger (1966). In figure 6.29 the simplest arrangement of a gamma camera in coincidence with a single scintillator is shown. It may easily be appreciated how this arrangement (figure 6.29(*a*)) is the nuclear medicine analogy of simple X-ray imaging (figure 6.29(*b*)). In the latter, of course, all X-rays emanate

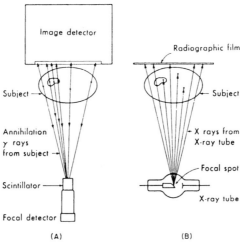

Figure 6.29 (*a*) A positron camera (with no tomographic facility) with single remote gamma counter as the focal detector. (*b*) Apparatus for taking a planar X-ray image of the same object. Note the geometry of image formation is identical in the two cases. (From Anger (1966).)

from a point and strike the object, and the attenuated beam is recorded at the film. In the former the gamma rays emanate from the patient but all gamma-ray pairs registered at the camera must have travelled along straight lines which connect to the scintillator; hence the geometry is identical. This then is not a tomographic imaging arrangement for positron emission.

† In fact it was a positive hydrogen ion; the positron was discovered in 1932 by Anderson (Feindel and Yamamoto 1978).

To obtain tomographic images more than one focal detector is required and possible arrangements are shown in figure 6.30. That utilizing a multielement focal detector is described in detail below. The plane sharply in focus lies parallel to the camera and can be varied in position by appropriate electronic shifts in the signals.

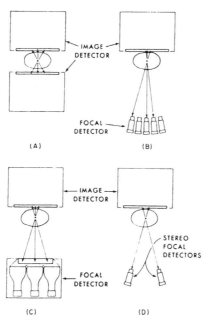

Figure 6.30 Four different arrangements whereby positron tomographic images may be obtained using a gamma camera. (*a*) Two gamma cameras are placed in close proximity to the patient. (*b*) A focused positron camera employing a gamma camera and an opposing set of focused scintillation counters. (*c*) Focused positron camera with remote solid-crystal focal detector. An array of photomultiplier tubes determines the position of the scintillations in the crystal. A relatively thick crystal is used to increase efficiency. (*d*) Stereo positron camera with two focal detectors. (From Anger (1966).)

Some of the earliest work in this respect was done by Anger himself at Berkeley. A gamma camera was set up opposite a so-called 'positron focal detector' comprising 19 scintillation counters arranged in a hexagonal array. A positron interaction was detected by the usual coincident methods with single-photon detection consequently eliminated. The neat trick was to make use of the information where one photon hit the focal plane detector to rearrange the recorded data on the collimator-less gamma camera. The interaction site was shifted an appropriate amount depending on which of the 19 crystals recorded the other 511 keV gamma. For a given strength of correction signal, one single plane between the two detectors was preferentially in focus and the position of this plane was varied by varying the correction signals. This was in effect a form of 'electronic seriescopy' and is somewhat reminiscent of similar methods developed by Muehllehner and McAfee for single-photon longitudinal-section tomography. The use of a bank of 19 detectors also increased the sensitivity as well as providing resolution in the plane in focus (Anger 1963). Anger used the positron camera to image midline brain tumours with ^{68}Ga ethylenediaminetetraacetic acid and functioning bone marrow with ^{52}Fe.

Similar developments were made by Kenny *et al* (1968) and Kenny (1971) at Sloan Kettering Institute. The collimation was removed from a facing pair of Anger cameras and they were set up in coincidence for imaging positrons.

Investigating this problem came to prove quite popular since it could make use of relatively cheap and widely available hardware, the gamma camera. In the mid-1970s a group in Chicago tried two techniques for positron emission tomography with a pair of facing gamma cameras (Muehllehner *et al* 1977). In both cases the cameras had their collimation removed. In the first case the cameras were kept static and sets of planes were reconstructed parallel to the faces of the detectors. This longitudinal reconstruction mode (so called because the planes were parallel to the detectors) suffered (as in all such methods) from cross-plane interference,

Figure 6.31 The positron camera showing rotational stand with opposing Anger cameras and control electronics. (From Muehllehner *et al* (1977).)

and a mathematical technique of deconvolving the blurring function was developed to cope with this problem. The advantage of this form of imaging was its relatively high sensitivity with useful images obtainable with of the order of 500k counts. In contrast with earlier developments along similar lines the sensitivity was also enhanced by using thicker crystals (1 in) and implementing pulse-shortening techniques and low-dead-time electronics.

The positron camera was also operated in rotational mode in which a form of cone-beam tomography was carried out. So great were the number of possible coincidence channels that data were collected in what is now called list mode (event by event) and then back-projected to form blurred tomograms. The blurring function was deconvolved from the back-projected data sets. This is the method still used in conjunction with rotating area detectors. The mechanism for rotation can be seen in figure 6.31 and useful clinical images were obtained with this arrangement.

6.5 Summary

The period of development of emission CT overlapped with that in which transmission CT was making great strides. Not surprisingly the two fields interacted. For example much of the complex mathematics of recon-struction from projections served both purposes. The two fields whilst engaged in some friendly rivalry for solving clinical problems comple-mented each other by measuring different physical properties of biological tissue. It is possible to identify where techniques from classical X-ray tomography were beaten into service for emission imaging. Because of the ready availability of digital detectors and fast computers, many of the problems inherent in X-ray longitudinal tomography did not reappear when longitudinal-section imaging was attempted for locating the distri-bution of radiopharmaceuticals. Other problems, however, replaced them such as worrying about the inherent lack of sensitivity with realistic doses, photon attenuation in tissue and poor spatial resolution.

The devices which were engineered matched these problems in ingenuity and, although many of the pieces of equipment recalled here are now of no more than historical importance, they can be viewed as the connecting link between the influences of classical X-ray tomography and the modern practices of SPECT and PET.

Epilogue:

Why Tomography?

The creation of images of the internal structures of the body, both normal and pathological, has resulted from applying the principles of physics and engineering to medicine. What factors have influenced the development of body-section imaging or tomography? As various developments from the origins of tomography have been addressed here, this question has in some measure been answered.

First and foremost has been the driving force of clinical motivation. Had planar X-radiographs been satisfactory, doubtless there would have been no tomography but, within a year of the discovery of X-radiation, the need for three-dimensional imaging had been voiced and was beginning to be part satisfied through stereoscopic radiology. The growing realization that some diagnoses were being inadequately made fuelled early experiments in blurring tomography; of particular interest were the problems of imaging the bony structures of skull and vertebral column, soon followed by the urge to improve pulmonary imaging, particularly where fluid-filled cavities were involved. Investigating the structures of the brain was a powerful motivation in the early days of X-ray CT. It was uppermost in the minds of both Oldendorf and Hounsfield. CT replaced other clinical examinations which were invasive, associated with trauma, discomfort and a high risk of complication or even death. In the War years the problems of precisely locating embedded projectiles escalated and were often solved by developments in tomography. Clinical demand also fired tomographic imaging in nuclear medicine for the assessment and quantitation of physiological function, diagnosis based on the abnormal uptake of a radiopharmaceutical by pathological tissues and the desire to study metabolic processes using radiotracers.

Much in the origins of tomography was synergistic. To a great extent, developments were dependent on improvements in ancillary equipment. Non-computed (blurring) tomography relied on moving X-ray sources and this in turn made demands on the stability and robustness of shielded tubes. In the early days those apparatuses in which tube movement was avoided, by requiring patient and detector to move instead, were often preferred for this reason. X-ray CT also relied on the ability to accelerate and decelerate X-ray sources safely and rapidly, required special tubes to deliver a high photon fluence, had to wait for the availability of efficient photon detectors and, perhaps most important of all, was quite impossible as a routine examination until the ready availability of the high-speed digital computer. Although there have been great developments in image reconstruction theory since the time when clinical CT became possible, the problem of what to do with the data once acquired had been essentially solved decades earlier. The tomographic mathematics was largely waiting on the solution of the data capture problem and a means for fast reconstruction. In its modern form CT started in the late 1960s but, prior to this, analogue data capture and section reconstruction had already been achieved. The influence of industrial interest, combined with the out-standing difficulty which still remained at this time of imaging the structures of the brain, accelerated the technique once the data capture problems had been solved and the computational requirements had been met.

Body-section imaging in nuclear medicine also developed as simul-taneously progress was made in several contributing areas. The ready availability of suitable radionuclides, satisfactory labelling of radiopharma-ceuticals, progress in detector technology and fast digital computers all combined to make emission tomography possible. Added to these were the clinical pressures to use tracers for uptake and metabolic studies, the cross-fertilization of ideas from X-ray imaging and an existing body of knowledge on reconstruction. The invention of the Anger or gamma camera played a vital role. It was used as the basis of both longitudinal and transaxial imaging methods and for positron tomography. Although the earliest section-imaging techniques used instead one or more moving single-crystal detectors, and indeed these developments featured large in the story of the origins of emission tomography, it is to the gamma camera that most modern SPECT developments have returned for their basis and, although non-camera-based equipment is still being developed, its supporters are in the minority. For PET the camera has proved less useful than might have been expected from the early work and multiple-detector systems or non-scintillator-based area detectors are now the methods of choice. Emission CT and X-ray CT have mutually benefited each other. First, emission CT seemed to be technically more advanced. It was soon overtaken by transmission CT and only now is emission CT catching up technically.

Of course these two are not rivals since they measure quite different properties of the human body and have separate roles to play. Both these post-Second World War developments have of course been simultaneous with the introduction of diagnostic ultrasound and again, whilst this modality also measures quite different physical properties of tissue, the spur provided by competitive and complementary examinations by ultrasound should not be underestimated. Possibly the development of ultrasound imaging was hindered since much of the earliest activity focused on the head, the clinical problem where it still fares worst. Fortunately X-ray imaging and emission imaging are conversely easier for the head than for the body and today section imaging with X-ray CT, with emission CT and with diagnostic ultrasound play complementary roles alongside each other.

It is possible to identify some 'push–pull' factors influencing body-section imaging. Firstly developments spanned two World Wars which stimulated initiative towards locating embedded projectiles and towards cheaper systems for tomography. Tailoring the suitability of equipment for field hospitals and particularly overcoming the problems of lack of portability of some cumbersome tomographic apparatus led to several innovations in design. Some doctors encountered clinical problems they would otherwise not have seen and certainly had to cope with more difficult circumstances. The Second World War accelerated existing interest in low-cost systems and simple equipment. It might be argued that the development of radionuclides also received a boost. At the same time, of course, war restricted supplies and taxed ingenuity. It certainly affected the flow of scientific information between European countries and also worldwide. The UK had to develop alternatives to German tomographic equipment. Several inventions were made several times over in differing parts of the world. Access to scientific literature was more restricted.

A second 'push–pull' factor was the influence of industrial interest. Without a doubt some of the earliest pioneers of section imaging found industry unresponsive to their requests and the existence of patents hindered rather than helped matters. On the other hand the role of industry in pioneering both transmission CT and emission CT was of paramount importance. This reflects the fact that the modern history of tomography has evolved on a quite different financial basis. CT has always been expensive. Even when the computers on which it relies have become cheaper, the hardware-costs of detectors, X-ray sources of the required quality, gantries, etc, have remained very expensive. The job has been one for industry and not the hospital or home workshop which coped so well in the days of Ziedses des Plantes and Watson. Fortunately public awareness of the importance of tomography has been such that Government and charitable funding has been (relatively) forthcoming in the UK. In the USA the situation has been even better, possibly on account of the differing funding methods and the importance placed on avoiding litigation in the case of diagnostic error. Scientifically the pressure to use tomography to assist planning and monitoring the effects of radiotherapy and to assist drug development with physiological imaging have only served to fan the industrial developmental fires.

Classical tomography has been largely the product of inspired individuals rather than collective groups. Many papers were from single workers, presenting equipment from their own workshops. We know some of the workers crafted the apparatus themselves. They were sometimes in ignorance of the work of others and papers often made no references. Consultants and radiographers built equipment. Tomography was not the exclusive domain of physicists and engineers. Often these clinicians were fighting apparent prejudice. It was written for example (Ziedses des Plantes 1973) of this worker that 'he expressed himself very carefully because at that time (1933) the general idea still was that planigraphy had no or little practical value'. It is not easy to establish how widespread was reactionary medical prejudice but one must suspect that many who prided their ability to diagnose from planar films might not entirely welcome newer ideas.

In the history of tomography there have been some remarkable coincidences. The presentations by Ziedses des Plantes and Bartelink of the planigraphic technique at that same meeting in 1931 comes immediately to mind. A much later coincidence was that of the two presentations of rotating-collimator emission tomography by Muehllehner and Freedman in 1970.

There have also been some grand confusions, not least associated with the words used to describe section imaging. These led at worst to complete misunderstandings of what some people were proposing.

There are other mysteries, such as why some patents were so long unknown (for example that of Bocage) or possibly ignored (for example those of Watson).

This book is not intended to be a complete and comprehensive review of all tomography. It has concentrated on the earliest origins of tomography and brought these to the point where a modern worker might begin to recognize familiar techniques and equipment. No attempt is made to review in depth modern developments which have stemmed from these origins. These are well covered elsewhere in review articles. There have been generations of X-ray CT machines, revolutions in detector development, computing, reconstruction and display. Ultrafast CT is possible. Elastic scatter tomography, charged-particle tomography, megavoltage tomography, simulator-based tomography have all been proposed and are being actively worked on. SPECT has moved through many special-purpose single-section imagers, to multiheaded gamma-camera systems. Positron tomography has become very expensive and multiring systems now play a major role in physiological measurement. On the tail of ionizing-radiation tomography have come ultrasound reconstruction tomography, magnetic resonance imaging and applied potential tomography. Much of this development relies on the digital computer and, if one dreams, what would one dream of the possibilities open to the early tomographers of the film-as-detector era if they had had at their disposal modern detectors and such computing tools?

Modern imaging of sections of the body has a long and colourful history. This has been a story of pioneering individuals. We are indeed standing today 'on the shoulders of giants'.

Appendix 1

Important Developments in X-Radiology and Nuclear Medicine for Contextual Framing of Tomography

The history of radiological imaging has attracted many writers and there is a wealth of literature describing the development of physical X-ray apparatus and its use in medicine. The chronicles of these developments together with the growth of radiological societies, radiological journals, manufacturing companies and the profiles of the pioneers are legion. The list which follows merely gives a contextual flavour, the background as it were, to the story of tomography. Not all the attributions are entirely indisputable and those mindful to know more might look at books by Brecher and Brecher (1969), Burrows (1986), Bruwer (1964), Allen *et al* (1966) and Grigg (1965). The latter in particular is surely unlikely to be surpassed as one man's account of radiological history. Even if a 'modern Grigg' could be compiled, would any publisher take the risk? Its encyclo-paedic proportions (26 short of 1000 large pages!) includes an anotated list of the radiological history literature up to its date (Felson 1977).

In the list which follows, however, some landmarks have been singled out using these and other sources. These are the discoveries and inventions which significantly advanced the course of radiology rather than the multitude of week-by-week developments. They took the physics of radiology through the pioneering year (1896), the gas tube era (1897–1913), the 'golden age of radiology' (1913–1945) into the atomic era (post-1945).

Date	Development
1874	The first Crookes (gas vacuum) tubes (as used by Röntgen)
1887	Goodman patented flexible (optical) photographic roll film
1888	Eastman marketed the first box camera (the Kodak)
1895	At the time of Röntgen's discovery the following types of high-voltage supply were available: Rühmkorff induction coil, Wimshurst influence machine, Thompson and Lemp X-ray transformer, dynamo with condensers
1896	Salvani 'cryptoscope'—a light opaque screen coated with barium platinocyanide (early fluoroscope)
1896	Edison discovered that calcium tungstate could replace barium platinocyanide as a fluoroscopic detector
1896	The idea of intensifying screens first mooted (Campbell Swinton in the UK and Pupin in the USA)
1896	Filters and diaphragms applied to Crookes tubes
1896	Probably the first X-ray department in the UK established at the Miller Hospital, Greenwich; others such as at the London, St Bartholomew's, The Royal Free and St Thomas soon followed
1896	Becquerel discovered radioactivity
1896	First reports of X-ray epilation
1896	The idea for a rotating-anode tube suggested by R W Wood
1897	The first full-body radiograph on a single film by W J Morton
1897	The (UK) Röntgen Society formed by Sylvanus Thompson
1898	Wooden radiographic couches (for example by Payne, Glew and Gardiner)
1898	First lead rubber gloves
1898	The Curies discovered polonium and radium
1899	First brain tumour imaged by X-rays reported
1900	Ilford Limited founded; manufacture of glass plates for X-ray work
1900	The Röntgen Society of the USA formed
1902	X-ray fluoroscopy (for localization) popularized by Shenton (Guy's Hospital)—the Shenton couch with undercouch X-ray tube and overcouch screen
1902	The American Röntgen Ray Society formed
1903	Allegemeine Electrizitats Gesellschaft (AEG) manufactured a stereofluoroscope with a twin coil and tubes
1903	Mackenzie Davidson's radiographic couch
1905	At the Berlin First Röntgenkongress the prefix röntgen- was adopted officially (for example Röntgenologie and the verb röntgenisieren) (the Viennese and Italians preferred the prefix 'radio-')
1905	The Pfahler filter (a piece of metal to filter soft X-rays) introduced to protect the patient
1907	Shenton introduced upright screening

Date	Development
1907	Tantalum targets introduced into gas tubes for higher current (50 mA) (Siemens)
1907	Snook first commercial transformer for high-voltage generation—the interrupterless transformer
1907	X-ray protective suits manufactured by Friedländer
1910	The curie defined as the unit of radioactivity
1911	First (US) professor of radiology (G E Pfahler at Philadelphia)
1911	Sir Archibald Reid introduced the overcouch tube at King's College Hospital, heralding a revolution in radiographic technique
1913	The Coolidge (hot-cathode–vacuum) X-ray tube (from William Coolidge 1873–1975)
1913	End of the 'gas tube era'; the 'golden age of radiology' began
1913	Salomon published the first mammograms
1914	Supply of continental glass for X-ray plates halted; search for replacement intensified
1914	Gustave Bucky invented the radiographic grid; War interrupted its introduction
1914	Eastman–Kodak marketed single-coated film
1915	The Dushman hot-cathode rectifying valve
1915	Potter developed a radiographic grid with movement to avoid unwanted shadows of grid wires
1917	Eastman–Kodak marketed double-coated film
1917	The British Association of Radiology and Physiotherapy formed
1917	Platinum in short supply in the USA for Coolidge tubes and tungsten substituted
1918	Introduction of air into the ventricles of the brain by Dandy (ventriculography)
1919	Dandy performs pneumoencephalography
1919	First oil-immersion X-ray tube–generator combination introduced by Waite
1919	Soluble iodine contrast agents first used
1920	The first radiological examination procedure (Cambridge Diploma) established in the UK
1920	The Radiological Society of North America formed
1920	The (UK) Society of Radiographers established
1920	The first commercial Bucky–Potter grid by G Brady Co
1921	Bucky grid manufactured in the USA and introduced into the UK
1921	Lipiodol (iodine containing oil) first used in France as contrast agent
1922	First radiological examinations by the American Registry of Radiologic Technicians
1923	Wilhelm Röntgen died
1923	Eastman–Kodak marketed 16 mm cine film

Date	Development
1924	Self-protected (Metalix) tubes incorporating chrome—iron fused to glass, a lead jacket and an exit window
1924	The British Institute of Radiology formed (incorporating the British Association of Radiology and Physiotherapy)
1924	Cellulose acetate 'safety film' produced (Eastman—Kodak) (Grigg records that this type of film had been made as early as 1889)
1925	The Victoreen condenser dosimeter was invented
1926	The Pohl Omniscope table for biplanar imaging
1927	Moniz performed cerebral angiography for the first time
1928	Safety film first produced in the UK at Harrow
1928	Barium sulphate suspension pioneered by GE for contrast studies
1928	The röntgen defined as the unit of radiation exposure
1928	Dye introduced first automatic mechanized film processor
1929	Rotating-anode tubes (the Rotalix)
1929	The first shock-proof tubes from Waite (for dental use)
1930	Hard glass bulbs replaced soda—lime glass for X-ray tube manufacture (Westinghouse)
1931	American Medical Association Directory lists 1005 radiologists in the USA
1931	Lawrence invented the cyclotron
1932	High-speed condenser discharge units introduced by Westinghouse enabled 'heart-arresting' radiography
1932	Discovery of the neutron
1932	Discovery of the positron
1934	Artificial radioisotopes discovered (transmutation)
1934	Marie Curie died
1934	Langmuir of General Electric patented the image intensifier (implementation delayed by the Second World War)
1934	Shock-proof oil-immersed portable X-ray tube produced by General Electric for (US) military use
1934	Picker produced an early biplane fluoroscope
1935	Double-focus X-ray tubes introduced (Westinghouse)
1935	Copper anode X-ray tubes introduced (Westinghouse)
1936	The Lawrence cyclotron capable of making at least 18 radioisotopes of biologically significant elements
1936	The Martyr's Memorial erected in the grounds of St George's Hospital, Hamburg, containing the names of (originally) 169 workers who died from radiation-induced disease
1937	Xeroradiography patented by Carlson
1938	$^{99}Tc^m$ first produced artificially with the cyclotron
1938	American Medical Association Directory lists 2191 radiologists in the USA (of which 10% practised diagnostic radiology only,

Date	Development
	5% therapeutic radiology alone, and 85% both; their mean age was 49 years; 95% worked in hospitals; 73% also worked privately—of these 82% alone and 18% in partnerships)
1938	Forced-oil-cooling systems introduced for X-ray tubes
1939	Flash X-ray photography possible with exposures of less than 0.1 μs
1942	Morgan invented the phototimer based on a fluorescent screen viewed by a photomultiplier tube
1942	Manhatten Project
1944	The improved Morgan–Hodges phototimer with variable 'stops'
1945	A stained glass 'Röntgen window', commissioned by Philips was executed by Joep Nicolas
1945	The Schmidt telescope principle applied by Bouwers (Oldelca) to create the photofluorographic mirror camera
1945	First atomic bombs
1945	At the end of the War there were 1336 radiologists in the US army
1946	Radioactive isotopes for medicine first shipped from the USA to the UK
1946	Oak Ridge reactor-produced isotopes available
1947	UK Harwell reactor first manufactured medical isotopes
1948	Westinghouse built the first image intensifier tube designed by Coltman
1948	First point-by-point isotope 'scan' (by hand)
1949	Chamberlain and Seal produce the first television chain (Orthicon) coupled to a fluoroscopic image
1950	Ziedses des Plantes, Cassen and Mayneord all working on scintiscanning independently
1952	Fluoroscopic image intensifiers were produced commercially (the Westinghouse Fluorex); combined with television, daylight viewing became possible
1952	Anger announced a crystal-plus-film gamma camera
1953	Estimated 125 000 X-ray sets in the USA operated by 80 000 technicians
1954	Society of Nuclear Medicine formed
1956	With the Eastman–Kodak X-Omat, automatic film processing came of age
1957	Anger's first all-electronic gamma camera
1958	$^{99}Tc^m$ first produced from a generator
1961	American Medical Association lists 7327 *full-time* radiologists

'The farther backward you can look, the farther forward you are likely to see.' Sir Winston Churchill (see Scott 1960)

Appendix 2

Glossary of Technical Terms

Most of the terms used in the text should be familiar to radiologists and scientists applying physics to medicine. Other readers may wish to consult a glossary of terms. As well as simple definitions, this appendix includes some descriptions of physical imaging techniques in enough detail to aid an understanding of the historical account. Refer instead to table 3.1 for the names of specific imaging equipment.

ANALOGUE RECONSTRUCTION Technique whereby section images are constructed without the use of a digital computer (for example by an optical method).

ANGER CAMERA Gamma-ray detector, named after its inventor, in which a large-area scintillation crystal is viewed by a matrix of photomultiplier tubes with electronics to give position sensitivity.

ANTIDIFFUSION STEREORADIOGRAPHY Stereo imaging of small regions with reduced scatter (after Chausse).

ANTISCATTER GRID Mesh usually of lead designed to enable passage only of primary (unscattered) X-rays and to reject other scattered X-rays.

ARTEFACT An unwanted structure in an image generally due to some unsatisfactory (but often insurmountable) feature in the imaging arrangement.

AXIAL TRANSVERSE LAMINOGRAPHY Vallebona's term for transaxial tomography.

BACK-PROJECTION Part of the CT reconstruction procedure. The values of data in projection elements are added to all the line-of-sight pixels in the reconstruction space. Also used for the name of the image formed by the above process when unfiltered projection data are utilized.

BETA EMISSION Electrons or positrons from a radionuclide.

BIOLOGICAL FUNCTION Term used to distinguish the properties of biological tissue which relate to how the tissue performs (metabolism, uptake of radiopharmaceutical, etc) rather than how it appears to attenuate X-radiation.

BIOLOGICAL TISSUES Normal and pathological tissues including organs within the body of a patient.

BLURRING TOMOGRAPHY A somewhat loose term for non-computed or classical tomography.

284

BODY SCANNER Rather loose term usually referring to a machine for CT of the body (rather than the head).

BODY-SECTION RADIOGRAPHY Term used by Moore for laminography; also generic modern term equivalent to tomography. (See also Tomography.)

BOOK CASSETTE Device for holding more than one X-ray film for simultaneous exposure.

CASSETTE Device for holding X-ray film.

CENTRAL-SECTION THEOREM Theorem which states that the one-dimensional Fourier transforms of projections of some two-dimensional distribution are corresponding central diameters in the two-dimensional Fourier transform of that two-dimensional distribution. (This theorem provides the link between Fourier CT methods and 'real-space' or convolution and back-projection CT methods.)

CHOPPER A mechanical device which acts as a time-variable shutter to an X-ray source.

CIRCULAR PLANIGRAPHY Planigraphy wherein the X-ray source and film execute circular movements.

CLASSICAL TOMOGRAPHY Generic term for all non-computed or so-called 'blurring tomography'.

CLOSE PACKING Requirement for individual radiation detectors in CT systems not to have large gaps between them.

COINCIDENCE DETECTION Form of imaging wherein two gamma rays are detected simultaneously (or nearly so). The gammas may be those from a positron-emitting radionuclide or an isotope emitting two gammas in cascade.

COMPOUND SCANNERS Apparatus for performing CT in which the X-ray source and/or detector execute complicated movements. The geometry of data capture does not fall into one of the usually accepted classes or 'generations'. Also term describing certain emission focal-plane scanners in which a slant-hole collimator, attached to a gamma camera, and the table supporting the patient both rotate.

COMPUTED TOMOGRAPHY (CT) Body-section imaging in which the required image must be reconstructed from projection measurements usually using a digital computer. (See also Projection, and Reconstruction algorithm.)

CONFORMAL THERAPY In the old Japanese definition, term for rotation radiotherapy; in the modern definition, term for radiotherapy with the high dose volume tailored to the tumour.

CONVOLUTION AND BACK-PROJECTION Mathematical technique for reconstructing CT data from (filtered or convolved) projections.

COOLIDGE TUBE Form of X-ray tube named after its inventor and largely replacing 'gas tubes' after 1913.

CORONAL SLICE Tomographic section bordered by superior, inferior, right and left of the patient (that is face-on view); generally not obtained by direct reconstruction but by reorganizing transaxial sections.

CROSS-THREAD METHOD Technique of using a stereoröntgenometer for making measurements from stereo X-radiographs; devised by Mackenzie Davidson.

CURVED-PLANE TOMOGRAPHY Form of tomography, based on slit radiography, whereby a whole-cranium tomogram may be recorded; also known as pantomography. (See also Pantomography.)

CYCLOTRON Circular particle accelerator which may be used for the production of certain radionuclides.

DÉFORMATIONS PÉRIPHÉRIQUES Artefacts in stratigraphy away from the axis of rotation.

DELAY LINE READ-OUT Electronic technique whereby the multiwire proportional chamber obtains its position sensitivity.

DEMODULATION Process of inverting or removing a modulation (sometimes of an X-ray intensity).

DIGITAL RADIOLOGY Form of radiology (tomographic or non-tomographic) in which the image is captured (often by discrete detectors) and stored as a digital matrix of numbers.

DIVERGENT-BEAM COMPUTED TOMOGRAPHY Also known as fan-beam CT. (See also Fan-beam tomography.)

DRIFT Term having several meanings including the variable response of an X-ray detector over a period of time.

ELECTROFLUOROTOMOGRAPHY Form of transverse axial tomography wherein the detector is a fluoroscopic screen.

ELECTRONIC REREGISTRATION Form of data processing in emission tomography wherein, by combining the signals from detectors at different spatial locations, information in selective planes can be reinforced with information from other planes blurred out.

EMISSION COMPUTED TOMOGRAPHY Emission tomography requiring the use of a digital computer to reconstruct digital tomograms. When based on the detection of single photons, the technique is known as single-photon emission computed tomography (SPECT). When based on the coincidence detection of high-energy photons from positron emission, the technique is known as positron emission tomography (PET). (See also Positron emission tomography.)

EMISSION TOMOGRAPHY Body-section imaging of the distribution of uptake of a radiopharmaceutical or some other property of its metabolism.

FAN-BEAM TOMOGRAPHY Form of CT in which the X-radiation forms a wide-angle fan spanning the patient as used in third- and fourth-generation CT scanners.

FIELD OF VIEW Region in space which is viewed at all times during tomography.

FILM Term generally here meaning film sensitive to X-rays and not visible light.

FILTERED BACK-PROJECTION Technique (or image created thereby) by which CT data are reconstructed from filtered projections.

FIRST-GENERATION COMPUTED TOMOGRAPHY Form of rotate–translate CT using a single detector. (See also Rotate–translate computed tomography.)

FLUOROSCOPIC SCREEN Detector which emits light photons when X-ray photons are incident upon it.

FOCAL-PLANE TOMOGRAPHY Alternative name for longitudinal-section tomography.

FOCUSED-COLLIMATOR SCANNING Form of rectilinear scanning for imaging the distribution of a radiopharmaceutical using a detector to which is attached a focused collimator; the arrangement can give fair longitudinal (blurred) tomograms.

FOURIER TRANSFORM Mathematical transformation, named after its inventor, which converts data from a function of some variable (for example time) into the corresponding function of the inverse variable (for example frequency).

FOURTH-GENERATION COMPUTED TOMOGRAPHY Form of rotate-only CT in which the detectors comprise a stationary ring and the X-ray source only rotates.

FRESNEL ZONEPLATE Form of X-ray aperture which when attached to a gamma camera enables focal-plane tomography.

GAMMA CAMERA Term usually meaning Anger camera. (See also Anger camera.)

GANGED SET Two or more detectors in a CT scanner which share the same electronic components; only one operates at any one time.

GEIGER–MÜLLER DETECTOR Form of X-radiation detector.

GLASS-PLATE DETECTORS Form of radiation detector which pre-dated film.

HEAD SCANNER Rather loose term usually referring to a machine for CT of the head (rather than the body).

HOMOIOMORPHIC STEREOSCOPY Any form of X-ray stereo imaging which is not tautomorphic (see § 1.3.3).

IN-FOCUS PLANE That plane in classical tomography in which all points project to corresponding points on a detector which do not move about the detector during the exposure. The information (shadows) from other planes appears as a superposed blur.

INVERSE RADON TRANSFORM Mathematics embodied in reconstruction theory for CT.

ISOCENTRE Axis of rotation of a gantry, for example that of a CT scanner or a machine delivering radiotherapeutic X-radiation.

ISORESPONSE PROFILE Lines along which the sensitivity of a radiation scanner is approximately constant.

LAMINAR TOMOGRAPHY Same as focal-plane tomography generally applied to emission imaging. (See also Focal-plane tomography.)

LAMINOGRAPHY Name used by Kieffer for planigraphy.

LAYER FLUOROSCOPY Longitudinal-section imaging, with a circular blurring trajectory making use of a fluoroscopic screen instead of a photographic film or plate.

LIGHT PIPE Device for channelling optical photons, for example from a scintillation detector to a photomultiplier tube.

LIMITED-ANGLE DATA Projection data from a CT scanner in which the range of orientations of the detector relative to the patient is not a complete circle.

LONGITUDINAL COMPUTED TOMOGRAPHY Form of CT whereby body sections are formed parallel to the long axis of the body. Sometimes used to mean form of imaging whereby planes are reconstructed parallel to some planar detector irrespective of orientation in the body.

MODALITIES Techniques whereby medical images may be obtained by physical probes.

MONOCHROMATIC SOURCE X-ray source emitting rays of only one energy.

MULTICRYSTAL POSITRON EMISSION TOMOGRAPHY SCANNER Form of positron tomography device using many scintillation crystals usually arranged in a circle.

MULTILAYERED CASSETTE Form of X-ray film cassette able to hold several films, generally parallel, for the purpose of obtaining several body-section images simultaneously.

MULTIPLE-BEAM RADIOTHERAPY Radiotherapy of tumours by combining one or more X-ray beams from different directions relative to the patient.

MULTIPLE-FILM SLICE FOCUSING Techniques such as seriescopy whereby a number of films exposed with the source in different locations relative to the body are viewed together (with appropriate shifts) in order to reinforce structures in one plane at the expense of others. This is in contrast with classical tomography based on the use of single films per slice.

MULTIWIRE PROPORTIONAL CHAMBER Form of gamma-ray detector based on proportional chamber theory.

OBLIQUE TOMOGRAPHY Form of tomography wherein arrangements are made for the slice in sharp focus to lie in some plane other than transaxial, coronal or sagittal (invented by Watson).

OPTICAL COMPUTED TOMOGRAPHY Form of CT reconstruction technique which it was possible to perform long before digital computers enabled CT (such as that of Frank).

PANTOMOGRAPHY Form of tomography invented by Paatero for recording a sharp image of the outer structures of the cranium (and specifically teeth) like a flat map of the world.

PARALLEL-BEAM COMPUTED TOMOGRAPHY Form of CT in which all the ray paths to projection elements (at each orientation) are parallel (that is first-generation CT).

PARALLEL-HOLE COLLIMATOR Collimator for an Anger camera enabling SPECT in parallel-beam (as opposed to fan-beam or cone-beam) geometry.

PARTIAL VOLUME EFFECT Effect whereby the value of a physical parameter ascribed to a voxel is some average value due to the voxel spanning tissues of different types.

PENDULUM TOMOGRAPHY Tomography of the Grossmann type.

PENETRATING RADIATION Any radiation which at least partially passes through the body but more specifically X-rays.

PENUMBRA Part shadows in radiology caused by (among other things) finite source size and collimation.

PHANTOM Object generally comprising tissue substitute materials used to simulate a patient or part thereof.

PIXEL Picture element in a digital image matrix (for example a CT scan).

PLANEOGRAPHY Form of multiple-film radiography invented by Kaufman for viewing and making measurements in three dimensions; closely related to seriescopy.

PLANIGRAPHY Technique whereby body-section images are formed on film by simultaneously moving a source and film in parallel planes about a line or point fulcrum. The plane in the patient through such a fulcrum, and parallel to the planes of movement of film and source, remains sharply in focus whilst other planes are blurred. The method was invented by Bocage and Ziedses des Plantes independently in 1921.

PLESIOSECTIONAL TOMOGRAPHY Multiple-plane tomography.

PLURIDIRECTIONAL PLANIGRAPHY Planigraphy wherein the X-ray source and film execute linear, circular and spiral movements.

POLYTOMOGRAPHY Multiple-plane tomography.

POSITION-SENSITIVE DETECTOR A radiation detector in which the position of the X-ray interaction can be very precisely determined. Some position-sensitive detectors are multielement (for example crystal scintillators); other continuous detectors use position sensing electronics (for example as in the Anger or gamma camera).

Positron Emission Tomography (PET) Form of sectional imaging of a radiopharmaceutical in which the radionuclide is a positron emitter (in contrast with a single-gamma emitter).

Preferred Embodiment Particular method of applying patented claims.

Projection Sum along a particular (usually straight) line path through a patient of some physical property of biological tissue. In transmission CT the projections are sums of X-ray linear attenuation coefficient. In emission CT the sums are complicated functions of both the uptake of a radiopharmaceutical in body tissues and the photon attenuation properties of the biological tissue.

Projection Filtering A vital stage in the mathematical process of reconstructing cross-sectional images from projections (CT). After projection filtering, data are back-projected into the reconstruction space.

Proton Computed Tomography Form of CT using protons as probe.

Radiodensity Loose term for X-ray linear attenuation coefficient (which is related to the electron density of tissue).

Radionuclide Isotope of an element which is radioactive.

Radionuclide Scanning Technique of generating a (generally non-tomographic) map of the distribution of a radiotracer in a patient by a (generally rectilinear) traversing movement of a radiation detector.

Radiopharmaceutical Pharmaceutical, element or compound labelled with a radionuclide.

Radiostereometry Measurement of distances within a patient by means of stereo pairs of X-ray films.

Radiotherapy Treatment Fractions Delivery of part of a radiotherapy treatment via a series of visits for the patient.

Radiotomie Proposed early term (not widely adopted) for tomography.

Range Telescope Part of the detector used for charged-particle tomography.

Raster Scanning Form of movement (usually of an X-ray detector) which resembles the method of scanning used in television.

Ray Reordering Part of CT reconstruction method whereby fan-beam data are converted to the equivalent parallel beam data sets.

Reciprocating Shutter Mechanical device which by oscillation variably collimates an X-ray source or detector.

Reconstruction Algorithm Mathematical method by which an image may be created from measurements of some physical property of biological tissues (such as X-ray linear attenuation coefficient or the uptake of a radiopharmaceutical) along a large number of different paths through the patient.

Rectilinear Scanner Apparatus comprising a radiation detector which generally scans in a plane by a raster fashion.

Relaxed Iteration A form of CT reconstruction whereby the pixel values are repeatedly adjusted until the mathematical projections of the object so formed match the real projection data (in a least-squares sense).

Röntgenography Old term for radiology.

Röntgenologist Old word for radiologist and sometimes X-ray technician.

Röntgenoscopy Old term for fluoroscopic X-ray viewing

Röntgen Rays, Tube, etc Old term for X-rays, X-ray source, etc, now disused.

Rotate-Only Computed Tomography Form of CT in which an X-ray source delivering a fan beam of radiation onto a bank of detectors (the fan completely spanning the patient) rotated (with no translation) about the patient. If the source and detectors both rotate, then this is also known as third-generation CT scanning; if only the source rotates and the detectors comprise a stationary ring, this is known as fourth-generation CT.

Rotate–Translate Computed Tomography Form of CT in which a single X-ray source and single detector (or small bank of detectors) are translated in a straight line relative to a patient with subsequent rotation of the line of translation through a large number of orientations relative to the patient. The projections so recorded form the basis of reconstruction. The form of CT with a single detector is also known as first-generation CT; the form with a bank is known as second-generation CT. (See also Projection.)

Rotating-Collimator Tomography Form of emission focal-plane tomography in which a gamma camera fitted with a rotating slant-hole collimator is used.

Rotation Radiogram Takahashi's old term for the sinogram. (See also Sinogram.)

Rotation Radiography Takahashi's term for section imaging. Numerous forms were invented (see § 2.4 for details).

Sagittal Slice Tomographic section bordered by superior, inferior, anterior and posterior of the patient (that is side-on view); generally not obtained by direct reconstruction but by reorganizing transaxial sections.

Scatter Rejection Method of removing the unwanted contribution to a radiograph or tomographic section from secondary X-rays.

Scintillation Detector Form of X-ray detector in which X-ray energy is converted into optical photons.

Scout View A digital two-dimensional planar radiograph formed using a CT scanner operated such that the source and detector do not rotate about the patient who is translated relative to the X-ray beam (used for localization or patient set-up.

Screen–Film X-ray detector comprising both an intensifying screen and a film (opposite to non-screen–film).

Second-Generation Computed Tomography Form of rotate–translate CT using a bank of detectors. (See also Rotate–translate computed tomography.)

Section Imaging Tomography.

Semiconductor Detector Form of X-radiation detector.

Sensitive Point Scanning Method invented by Oldendorf whereby the internal structure of some complex organ could be determined even though that structure were deeply embedded in overlying tissue (for example detection of the ventricles in brain).

Seriescopy Use of four or more films taken from different locations in a plane of the X-ray source in a device called a Seriescope to enable the viewer to reinforce information in one plane at the expense of others and thus enable the viewer to view sequentially a whole series of planes through the patient as if by stepping through the patient. Invented by Ziedses des Plantes.

Serioscopy Alternative spelling for seriescopy. (See also Seriescopy.)

Shielded Tube X-ray source with radiation-protective shielding.

290

SINGLE-FILM SLICE FOCUSING Generic term for all tomography (as opposed to multiple-film seriescopy) using a single film per slice.

SINGLE-PHOTON EMISSION COMPUTED TOMOGRAPHY (SPECT) See Emission computed tomography.

SINOGRAM Image formed by stacking projections, one beneath the other, taken by detectors at sequential (and generally equal) angular increments around a patient. An off-axis point would describe a sine wave in the sinogram—hence its name. (See Appendix 4.)

SKIAGRAPHY Very old term for radiography.

SLANT-HOLE COLLIMATOR Collimator for gamma camera in which all the holes are parallel to a plane which is not normal to the face of the collimator.

SLICE THICKNESS Thickness of a tomographic section.

SLIT RADIOGRAPHY Form of radiography where only a narrow slit of radiation was used. Some tomographic techniques were based on this, for example pantomography.

SODIUM IODIDE DETECTOR Common form of X-radiation scintillation detector.

SOLIDOGRAPHY Creation of solid models of structures from outlines obtained from tomograms.

SOMERSAULTS OF ZIEDSES DES PLANTES Form of ventriculography named after its inventor whereby a small quantity of air is imaged sequentially in all four ventricles by arranging for the patient's head to take up the positions as in a somersault.

SPIRAL PLANIGRAPHY Form of planigraphy wherein the X-ray source and film execute spiral (rather than linear or circular) movements.

STEREORÖNTGENOMETER Apparatus whereby the geometry of stereo X-ray imaging is copied and strings or wires are used to trace ray paths for making measurements (for example of the bony pelvis).

STEREOSCOPY Real-time stereo viewing using a double-focus X-ray tube or two tubes and a fluorescent screen.

STEREO X-RAY IMAGING The use of a stereo pair of X-rays to obtain a three-dimensional effect; attributed to Elihu Thomson.

STRATIGRAPHY Technique whereby body-section images are formed on film by simultaneously moving a source and film at opposite ends of a pendulum in a rotation about a line fulcrum with the film remaining perpendicular to the line of the pendulum at all times. Strictly only the line of the fulcrum remains sharply in focus. The method was invented by Vallebona. The term also applies to equivalent arrangements whereby the patient was moved with respect to a stationary X-ray source and detector. The term was coined by Busi.

TAUTOMORPHIC STEREO IMAGING Stereoradiology performed under those conditions (see text) designed to yield accurate representations of the object being investigated in all spatial parameters.

THERAPY COMPUTED TOMOGRAPHY SCAN CT scan (for localization purposes) taken with the patient in the same position as they will be set up for subsequent radiotherapy.

THIRD-GENERATION COMPUTED TOMOGRAPHY See Rotate-only computed tomography.

TISSUE INHOMOGENEITY CORRECTIONS Part of the radiotherapy planning process which takes account of the varying X-ray attenuation of different body tissues. CT data are often used to assist this process.

TOMOGRAM Term used for the body section generated by tomography.

TOMOGRAPH Term sometimes used for the body section generated by tomography. (Also confusingly used for the equipment to generate the same.)

TOMOGRAPHY Strictly (and historically) the technique whereby body-section images are formed on film by simultaneously moving a source and film at opposite ends of a pendulum in a rotation about a line fulcrum but such that the film remains parallel to a plane through the fulcrum. Sharp images are obtained of only this plane, all other planes being blurred. The method was invented by Grossmann in 1934. Since 1962 the term has been decreed by ICRU to be applied to *all* methods of body-section imaging, howsoever performed.

TOMOSCOPY Section imaging by viewing a fluoroscopic screen.

TRAJECTORY Path taken by an X-ray tube or detector.

TRANSAXIAL COMPUTED TOMOGRAPHY Form of CT where sections are formed transaxial to the long axis of the patient's body.

TRANSAXIAL TOMOSCOPY Watson-type transverse axial imaging but with images viewed on a screen rather than recorded by film.

TRANSMISSION COMPUTED TOMOGRAPHY X-ray CT as for example distinguished from emission CT.

TRANSVERSE AXIAL TOMOGRAPHY or TRANSAXIAL TOMOGRAPHY Imaging whereby the rays from a stationary X-ray source are cast obliquely through a rotating patient onto a synchronously rotating detector. Invented by Watson in 1937. Not to be confused with transaxial computed tomography.

TRAVERSED FOCUS RADIOGRAPHY Term sometimes used for 'Mayer's method' (see §§ 1.2.11 and 3.2) whereby structure in contact with a film was shown sharpest as the X-ray set moved during the exposure.

TUNGSTATE SCREENS Form of intensifying screen for X-ray film.

UNFILTERED BACK-PROJECTION CT image formed from projections which have not been convolved with a filter.

VERTIGRAPHY Form of planigraphy in which the detector is perpendicular to the planes of motion of the X-ray source and the detector.

VOXELS Volume elements within the patient in which some physical property is measured and displayed.

WALSH–HADAMARD FUNCTIONS Specific arrangements of zeros and ones sometimes used to code the modulation of an X-ray source.

WATER BAG Bag of water placed around the patient in early CT scanners in order to reduce the dynamic range of the transmitted X-ray intensity.

WHEATSTONE STEREOSCOPE Apparatus for viewing stereo films, named after its inventor.

WINDOWS Term with many meanings in this field including openings in an X-ray shutter or aperture or a range of data displayed in a CT scan.

XENON DETECTOR Form of multiwire proportional chamber containing xenon gas; used in CT.

ZONOGRAPHY Tomography with very small angle of swing of the pendulum.

Appendix 3

From the Scientist's Art to the Doctor's Decision—the Evolution of Medical Display

It is all too easy for physicists to forget that their aim is to assist the medical doctor to understand what is happening to the patient. Towards this goal they provide images, pictures which for transmission X-radiology represent the varying X-ray attenuation properties of the differing tissues and for gamma emission represent their functional properties. Both transmission and emission radiology have been transformed by the development of tomographic techniques whereby the body may be viewed as if sectioned by a cleaver into slices. 'Classical' or 'blurring' tomography only partly achieved this representation and the effects of structure not in the plane of interest interfered with the sharp definition of what the clinician had hoped to see. CT has solved this problem.

The interface between the scientific method and the doctor's thoughts rests with the displayed image and, through the period of development of tomography, the means of display has also undergone many changes. The issue of optimizing display is still a contentious source of debate. Just as a poor presentation by a lecturer can fail to convey good material, so good diagnostic information can be lost through a poor method of image display.

The earliest tomographic images were recorded on films or viewed on fluoroscopic screens. Glass plates would also have been used and are mentioned in many early patents. These methods of recording possessed adequate dynamic range to accommodate tomographic images and it is unlikely that the pioneers of the early 1920s and 1930s felt limited in this respect. One exception was for seriescopy, the technique whereby different layers in the patient were brought into focus by superposing films taken with the X-ray source in differing locations relative to the patient. For this technique, some skill was required to adjust the intensities of the separate films so that the superposed set was not too dark. The same problem arose with multiple-plane tomography performed with several films being exposed simultaneously during one movement. These two techniques would both have been much more successful if digital computers had been available for then the images could have been enhanced and renormalized. (Electronic seriescopy was later developed for nuclear medicine—see chapter 6.) In some ways, tomographic images were considered (by the inexperienced) to be more rather than less blurred because they smeared out the 'usual' anatomical landmarks such as bony structure in order to sharpen the detail of interest. Figure A.1 shows what a 1940s radiologist might have been viewing (see also figure 1.37).

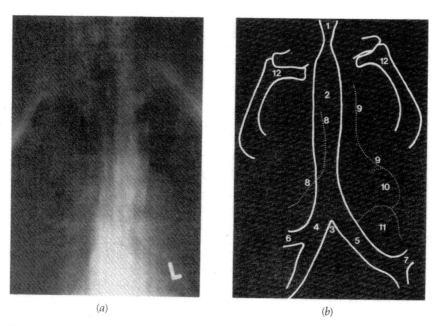

(a)   (b)

Figure A.1 (a) Anteroposterior 'classical' tomogram of the mediastinum showing the trachea and major bronchi without the shadows of other structures such as most ribs which would obscure the image. (b) A schematic diagram showing what can be seen on the tomogram: 1, larynx; 2, trachea; 3, carina (bifurcation of main bronchi; 4, right main bronchus; 5, left main bronchus; 6, right upper lobe bronchus; 7, left upper lobe bronchus (superior lobe); 8, pleural reflection of right upper lobe; 9, left pleural reflection; 10, aorta; 11, pulmonary artery; 12, first rib. (From Weir and Abrahams (1978).)

Transverse axial tomography was also film based in the implementations of Watson, Vallebona and Takahashi. The latter, in his experiments with what we should today call CT, provided some elegant hand drawings (see figure A.2) of cross-sectional data. Leonardo like, they relate the section to recognizable external anatomy (the reclining patient).

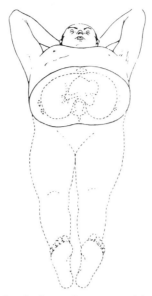

Figure A.2 One of the display techniques used by Takahashi to show transverse axial tomograms. (From Takahashi (1969).)

The early experiments of Gabriel Frank suffered from the absence of the proper tools for image reconstruction and display. Reconstruction was limited to back-projection by optical methods. Frank's method so nearly achieved CT but had the method been used the images would have been little better than transverse axial tomograms.

The development of nuclear medicine coincided with the 'new computer age' and, almost from the start, displays were based on cathode-ray tube display and computer-generated data. Not quite, for the first scans were made 'point by point' by hand movement of a simple detector such as a Geiger counter over the region of interest (Ansell and Rotblat 1948). The displays were also drawn by hand, simply mapping the recorded counts in space (figure A.3), non-tomographic of course. From these rude plots, bearing no resemblance to anatomical images, the clinician drew conclusions.

Automatic scanners and cameras with some tomographic properties also gave their results in a variety of suboptimal ways and these methods were also still in use when tomography became possible. The earliest images of positron emission were made by 'paper tappers' in which a different symbol was tapped depending on the position of the disinte-gration and the number of superposed symbols represented the intensity (figure A.4) (Brownell and Sweet 1953). In 'home-made' developments it was quite common for computer-generated images to be displayed as

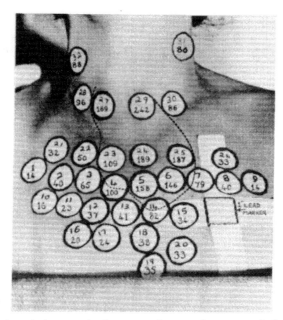

Figure A.3 A photograph of a patient showing the sites at which readings from a simple gamma counter were taken. The upper numbers show the order in which the readings were taken and the lower numbers are the background subtracted count. The broken line indicates the outline of the thyroid as calculated from the measurements of radioactivity. The lead marker is to allow comparison with a planar transmission X-radiograph. (From Ansell and Rotblat (1948).)

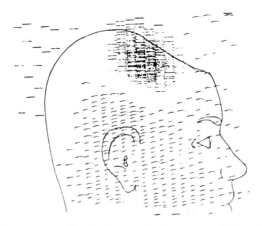

Figure A.4 An example of a 'two-symbol' computer display of a positron emitting radioactive distribution. The symbol indicates whether the decaying nuclide is in the right or left of the head and the intensity of the symbol is proportional to activity. (From Brownell and Sweet (1953).)

grey-scale overstrike' plots on paper until well into the 1970s (figure A.5). The earliest SPET machine (Kuhl and Edwards 1963) simply modulated the brightness of a spot on a cathode-ray tube with a line sweep introduced to effect the 'back-projection'. A hard copy could be made on film by an open aperture camera viewing the screen.

The real revolution in display coincided with the use of digital computers for image reconstruction and with the commercial launch of

Figure A.5 An image of an X-ray CT section displayed on a computer using 'overstriking'. Symbols are composites representing the X-ray attenuation with symbols appearing darker as attenuation increases. This shows a section through the thorax of a rabbit reconstructed with a simple 'first-generation' scanner built by the author and colleagues in the late 1970s.

X-ray CT. From 1972 both emission and transmission images were displayed as matrices of pixels on a television monitor, the brightness of each pixel being proportional to the magnitude thereof. A modern CT section is shown in figure A.6. This method of display is now so common as to be largely taken for granted. It permits grey-scale windowing and all those reorientation techniques whereby for volumes of reconstructed voxels (volume elements) the data may be displayed in sagittal, coronal and oblique as well as transaxial cross-sections. Additionally such data provide the basis for three-dimensional shaded surface display (figure A.7) and methods of dynamic display which it is not possible to illustrate in a book. The modern clinician has choices for reporting. Some prefer to sit at

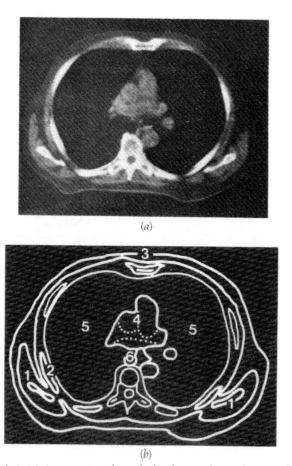

(a)

(b)

Figure A.6 (a) A CT section through the thorax. (b) A schematic diagram showing the main detail: 1, scapula; 2, ribs; 3, sternum; 4, mediastinum; 5, site of lung (structure windowed out in this display); 6, aorta. (From Weir and Abrahams (1978).)

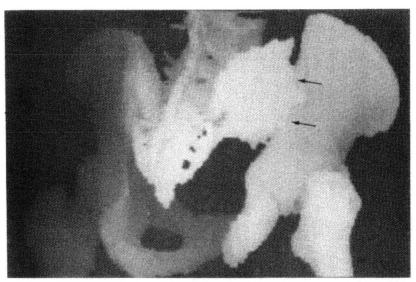

Figure A.7 Three-dimensional shaded surface display of the bones of the pelvis showing a fracture involving the right ilium and superior pubic ramus (illustrated) resulting in pelvic asymmetry (postero-oblique view). (From Gillespie and Isherwood (1986).)

the television monitor and view the data interactively. Others prefer to take hardcopy away and to use a variety of methods for this such as transferring the images to roll film, X-ray film, Polaroid film, etc. In a way a new problem has arisen that, when presenting data in this way (and also for publication), the subtleties one might be trying to highlight are lost in the modulation introduced by the printing process. In these days of rapid image manipulation it is sometimes possible to lose track of what are the raw experimental data. Perhaps avoiding this is the modern physicist's equivalent of the radiographer's skill in early days of choosing exposure conditions. It is a skill in which is vested much responsibility. Much is heard of the need to investigate the false positive and negative data which might ensue. This is a phenomenon remarkably absent in the history of early tomography. Wide choices in image manipulation and display simply did not exist. When choosing diagnostic imaging equipment today as much attention is paid to the ways of manipulating and displaying the data as in its acquisition. Now digital medical images are available from so many physical modalities, the hope for the future is that complementary information might be simultaneously displayed in some overlying manner. Although some have tried, very little real progress has been made on this and today we are accustomed to doctors' viewing separately the information from different quarters and performing the integrations in their mind.

Appendix 4

The Theory of X-ray Computed Tomography in Words and Pictures

The mathematics of X-ray CT is a rich field with a very large literature. This has built up to accommodate the many geometries of data capture, to develop methods of coping with noise, to reconstruct from irregular or incompletely sampled data, to achieve ultrarapid image formation and in response to a host of other problems in reconstruction from projections. The literature is of varying mathematical complexity but unfortunately most of it is beyond anyone who is not at home with the Fourier transform and convolution integrals. This greatly reduces the number of people to whom the physical basis of CT is readily apparent. It is, however, possible to present the basic ideas of reconstruction in words and pictures and this appendix is for those who feel this would help them to get more from the main text. The approach is not without limitations; not using mathematical symbolism generally makes the exposition longer and more clumsy in appearance. The discussion is limited to the method of 'convolution and back-projection' to which the main text has referred from time to time.

Imagine a single point absorber of radiation inside a circle of air (which is assumed not to attenuate the X-rays) (figure A.8(a)) and that this 'slice' is irradiated with a parallel beam of X-rays of width equal to the diameter of the circle. The beam is recorded by a one-dimensional detector with a large number of individual detecting elements. All the elements will detect the same X-ray intensity (let us ignore noise) except for the single central element which will detect a reduced intensity. The detector's response is shown to the right of the figure. From this, the total absorption along the line of sight of each detecting element can be obtained (as the logarithm of the ratio of the detected intensity for any beam missing the absorber to that in each element). Obviously this total line-of-sight absorption is zero except for the one central element which 'sees' the point absorber. The one-dimensional profile of line-of-sight absorption is called a 'projection' and this has also been shown on the extreme right of figure A.8(*a*). Now imagine the X-ray source and detector are rotated about the point absorber so that they take up a large number of regularly spaced orientations with respect to the point absorber. For the case where the only X-ray absorption is at the central point, all the projections are the same. (This would of course not be the case if some other object were in the field of view.) This observation of the attenuation in a slice 'edge on' from a number of orientations is the fundamental way in which CT scanners gather their data. They might do this, for example, by scanning a pencil of radiation, whose width is only one single detector element, across the field of view at each angular orientation. The problem is how to reconstruct the absorbing object from the set of one-dimensional projections.

Suppose the one-dimensional projections are laid out in their appropriate orientations with respect to the object and the value in each digital element of each projection is simply added back everywhere along the line of sight of that element. This is called back-projection. Wherever lines cross in the circular field of view, the back-projected values are added together at the crossing point. This will produce a spoked pattern of lines of non-zero values where all the lines cross at a point and there are as many spokes as there are one-dimensional projections. Figures A.8(*b*) and A.8(*c*) show what would be reconstructed from 16 and 32 projections respectively equispaced in the range 0–360°. This process correctly 'finds' the position of the point absorber (where the lines all cross). Unfortunately the spokes are unwanted reconstructed detail (artefact) arising from the back-projection. Now suppose the number of projections is made very large (and the angle between them very small), for example 3600 projections spaced at $1/10°$, then the spokes would merge to the viewer who would simply perceive that the pattern of reconstructed object was darker towards the crossing point and became less dark as the distance from that point increased. In this limit the reconstructed density would decrease in inverse proportion to the distance from the central point.

We could write

reconstructed signal at point P a distance from central crossing point A $\propto \dfrac{\text{reconstructed signal at point A}}{\text{distance of P from A}}$

Now suppose the projections are formed for a non-central point B. In this case they will all be different but, when they are laid out as before and

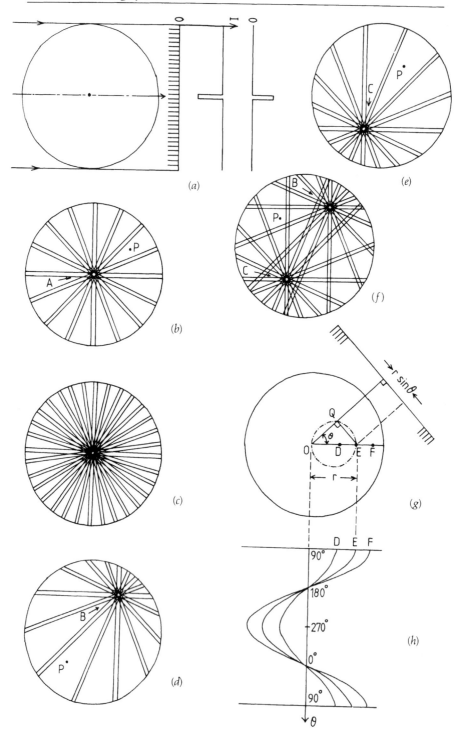

(a)

(b)

(c)

(d)

(e)

(f)

(g)

(h)

back-projected in the same way, the non-zero elements will generate spokes which will cross at the site B (figure A.8(*d*)). Again in the limit of a large number of projections being measured, we can write

reconstructed signal at point P a distance from crossing point B $\propto$ $\dfrac{\text{reconstructed signal at point B}}{\text{distance of P from B}}$

Now suppose there is a point absorber at a different non-central point C. The same procedure (shown in figure A.8(*e*)) gives a new spoke pattern centred on C and we may write

reconstructed signal at point P a distance from crossing point C $\propto$ $\dfrac{\text{reconstructed signal at point C}}{\text{distance of P from C}}$

Now consider what happens when *both* point absorbers B and C are present together in the X-ray field. In each projection, two non-zero elements are recorded. These will of course be different detector elements for each orientation of the detector about the object. When the projections are laid out and back-projected in the same way (each element 'connecting' to its line of sight in the reconstructed field), *two* spoke patterns would arise, one centred on B and the other on C (figure A.8(*f*)). This is an example of the concept of a 'linear system' meaning that the result of radiographing the two point objects together is exactly the same as adding together the pairs of projections at each angle of the separately radiographed single point absorbers. The back-projection is also a linear process and so in the limit of a very large number of projections we may then write

total reconstructed signal at point P $\propto$ $\dfrac{\text{reconstructed signal at point B in the absence of C}}{\text{distance of P from B}} + \dfrac{\text{reconstructed signal at point C in the absence of B}}{\text{distance of P from C}}.$

Figure A.8 (*a*) The formation of a projection of a central point absorber. (*b*) The operation of back-projection for just 16 projections of the central point absorber. (*c*) The operation of back-projection for just 32 projections of the central point absorber. (*d*) The operation of back-projection for just 16 projections of a non-central point absorber at B. (*e*) The operation of back-projection for just 16 projections of a non-central point absorber at C. (*f*) The operation of back-projection for just 16 projections of *two* non-central point absorbers at B and C. (*g*) The relationship between the elemental position in a projection and the position of a point in the section when the projection is at angle θ. (*h*) The sinogram corresponding to the three points D, E and F in (*h*).

Now it is a small step to recognize that an extended but discontinuous object is simply a sum of point absorbers and so we may write

$$\text{total reconstructed signal at P in an object} \propto \text{sum of} \left\{ \frac{\text{reconstructed signal at each point in the absence of all the others}}{\text{distance of P from that point}} \right\}$$

and in the limit as the object becomes continuous

$$\text{total reconstructed signal at P in an object} \propto \int \frac{\text{reconstructed signal at each point in the absence of all the others}}{\text{distance of P from that point}}$$

The left-hand side is the *back-projection*; the numerator of the right-hand side is proportional to the required *true object cross-section* and so we reach the well known statement (since the right-hand side is the definition of a convolution with a function inversely proportional to distance from each point)

$$\text{back-projection} \propto \text{true cross-section} * 1/r$$

where $*$ means 'convolved with'. The function $1/r$ is referred to as the 'blurring function' since it is the source of the degredation from true to measured cross-section if the reconstruction is by simple back-projection only.

We see then that the simple act of back-projection alone will not give a faithful map of X-ray linear attenuation coefficient in a slice. (Indeed this was the reason why transverse axial tomography as configured by Watson and others also generated blurred tomograms (they would have done so even for grazing-incidence X-rays).) To overcome this, something else needs to be added into the recipe for reconstruction.

To discover what else needs to be done, let us consider the picture which is formed by laying out all the one-dimensional projections in order of angular orientation one above the other. This is the diagram called the 'sinogram'. The vertical axis represents the angle made by the normal from the axis of rotation to the centre of each projection relative to the horizontal axis of the object and the horizontal axis in the sinogram labels the position within the projection. Thus the central vertical line, being the central element of the detector, represents the absorption along the line of sight passing through the axis of rotation. Consider now points D, E and F all lying in a straight line along the horizontal of the object but with increasing distance r from the axis of rotation (figure A.8(g)). From figure A.8(g) we see that, when the line, from the axis of rotation, normal to the detector is at angle θ to the horizontal each point projects to the element $r \sin \theta$ on the detector where r is the distance of the point from the axis of rotation. Triangles such as OQE are always right angled and thus the intersection of the line, from the axis, normal to the detector and the lines through D, E and F parallel to the detector (shown as point Q for point E) follow circles (of increasing radii) as the detector rotates about the object. (Only the circle corresponding to the intersection Q for the line through E is shown.) Hence the path of the appropriate detecting element through the sinogram (given by the distance EQ—for point E—as a function of θ) is simply a sine wave (hence of course the name sinogram). The paths are shown in figure A.8(h) for the three points shown in figure A.8(g). This provides the clue as to how to erase the unwanted spokes. Suppose the only absorber is once again the point at the centre O. The sinogram is then only non-zero along its central vertical stripe, but the path of the projection elements for the other points D, E, F in the circle pass twice per rotation through this line and thus receive an unwanted reconstructed density in proportion to the obliquity of the sine wave relative to the central vertical in the sinogram. This gives rise to the $1/r$ contribution to the off-axis points. The same would happen for all other points in the field of view except that the crossing points would not all coincide as they do for these three points in a line. The 'trick' is to multiply each projection by a filter which fills the rest of the sinogram with negative numbers such that the sum of the values along any sine-wave curve comes to zero. If this can be arranged for a point at the centre, from linearity the desired effect will be obtained for a more complicated object. In this case since the multiplying filter is applied centred on each detector element in turn, the operation becomes that of *convolving* the projections with the filter and the sinogram becomes filled with a distribution of positive and negative data. By doing this before back-projection the $1/r$ blurring function is effectively removed. The two operations combined give this method the name 'convolution and back-projection reconstruction'. The appropriate filter can be deduced from the mathematics and depending on the assumptions a variety of filters emerge. They can be tailored to achieve other desirable aims such as a reconstructed image with a smooth appearance.

The method simply reduces then to the following.

(1) Measure the one-dimensional projections for a full set of angular positions.

(2) Convolve these with a filter.

(3) Back-project the *filtered* projections along the line-of-sight of each detector element.

In this oversimple discussion the problems of normalization have been ignored; essentially the final image can be calibrated to give reconstructed values of linear attenuation coefficient.

References

ALEXANDER G H 1938 A simple and inexpensive tomographic method *Am. J. Röntgenol.* **39** 956–8

ALLEN K D A, MCFETRIDGE E M and STEIN M W 1966 *Radiology in World War 2* (Washington, DC: Office of the Surgeon General Department of the Army) pp 137, 269–71, 379, 389, 393, 461, 521

ALTSCHULER M D 1979 Reconstruction of the global-scale three-dimensional solar corona *Image Reconstruction Implementation and Applications* ed G T Herman (Berlin: Springer) pp 105–45

AMBROSE J 1977 CT scanning: a backward look *Semin. Röntgenol* **12** 7–11

AMBROSE J 1988 private communication (discussion, 11 November)

AMBROSE J and HOUNSFIELD G N 1972 Computerized transverse axial tomography *Br. J. Radiol.* **46** 148–9

AMBROSE J and HOUNSFIELD G N 1973 Computerized transverse axial scanning (tomography) Part 2: Clinical applications *Br. J. Radiol.* **46** 1023–47

ANDREWS J R 1936 Planigraphy 1: Introduction and history *Am. J. Röntgenol.* **36** 575–87

ANDREWS J R 1944 Röntgenography: body section *Medical Physics* (Year Book Publishers) pp 1264–7

ANDREWS J R and STAVA R J 1937 Planigraphy 2: Mathematical analyses of the methods, description of apparatus, and experimental proof *Am. J. Röntgenol.* **38** 145–51

ANGER H O 1958–61 Radiation image device *US Patent* 3011057

ANGER H O 1963 Gamma-ray and positron scintillation camera *Nucleonics* **21** 56–9

ANGER H O 1966 Radioisotope cameras *Instrumentation in Nuclear Medicine* ed G J Hine (New York: Academic) pp 485–552

ANGER H O 1968 Tomographic gamma-ray scanner with simultaneous readout of several planes *Fundamental Problems in Scanning* ed A Gottschalk and R N Beck (Springfield, Illinois: Charles C Thomas) pp 195–211

ANGER H O 1969 Multiplane tomographic gamma ray scanner *Medical Radioisotope Scintigraphy* (Vienna: International Atomic Energy Agency) pp 203–16

ANGER H O 1974a Multiplane tomographic scanner *Tomographic Imaging in Nuclear Medicine* ed G S Freedman (New York: Society of Nuclear Medicine) pp 2–18

ANGER H O 1974b Tomography and other depth discrimination techniques *Instrumentation in Nuclear Medicine* vol 2, ed G J Hine and J A Sorenson (New York: Academic) pp 61–100

ANGER H O, PRICE D C and YOST P E 1967 Transverse section tomography with the scintillation camera *J. Nucl. Med.* **8** 314

ANSELL G and ROTBLAT J 1948 Radioactive iodine as a diagnostic aid for intrathoracic goitre *Br. J. Radiol.* **21** 552–8

ANTOINE (initial unknown) 1952 La tomographie axiale transverse *Rev. Méd. Nancy* **77** 510–2

ARAUNER A 1982 Röntgenuntersuchungsgerat *Eur. Patent* 0074021

ARE 1919 Regulations and syllabus for the Cambridge Diploma in Medical Radiology and Electrology *Arch. Radiol. Electrotherapy* **24** 209–16

ARIMIZU N 1972 Method of tomographic imaging with different collimators *Medical Radioisotope Scintigraphy, Proc. Symp. Monte Carlo, October 1972.* (Vienna: International Atomic Energy Agency) pp 369–79

ARONOW S 1966 Positron scanning *Instrumentation in Nuclear Medicine* ed G J Hine (New York: Academic) pp 461–83

ARR 1911 (no title to article) *Arch. Röntgen Ray* **16** 228

ARR 1913 A new method of stereoscopic photography *Arch. Röntgen Ray* **18** 230–1

BAESE C 1915–7 Method of, and apparatus for, the localisation of foreign objects in, and the radiotherapeutic treatment of, the human body by the X-rays *UK Patent* 100491; *Ital. Patent* (several references quote 1915 but number unknown and search failed)

BAESE C 1917a Quelques remarques sur les localisations géométriques *Arch. Élec. Méd. Physiothérapie* **413** 59–69

BAESE (DI CASTELVECCHIO) C 1917b In tema di localizzazioni geometriche *Radiol. Med.* **4** 6–13

BAESE (DI CASTELVECCHIO) C 1918 A universal X-ray apparatus of great utility *Arch. Radiol. Electrotherapy* **23** 73–83

BAILY N A 1979 The fluoroscopic image as input data for computed tomography *IEEE Trans. Nucl. Sci.* **NS26** 2707–9

BAILY N A, CREPEAU R L and LASSER E C 1974 Fluoroscopic tomography *Invest. Radiol.* **9** 94–103

BAILY N A, LASSER E C and KELLER R A 1976 Tumor localisation and beam monitoring-electrofluorotomography *Med. Phys.* **3** 176–80

BARCLAY A E 1949 The old order changes *Br. J. Radiol.* **22** 300–8 (reprint 1950 *Yearbook of Radiology* ed F J Hodges, J F Holt, I Lampe and R S MacIntyre (Chicago, Illinois: Year Book Publishers) pp 10–7)

BARRETT H H 1987 private communication (letter to S Webb 13 October)

BARRETT H H, DeMeester A D, Wilson D T and Farmelant M H 1974 Tomographic imaging with a Fresnel zoneplate camera *Tomographic Imaging in Nuclear Medicine* ed G S Freedman (New York: Society of Nuclear Medicine) pp 106–21

BARRETT H H, HAWKINS W G and Joy M L G 1983 Historical note on computed tomography *Radiology* **147** 172

BARRETT H H and Swindell W 1977 Analog reconstruction methods for transaxial tomography *Proc. IEEE* **65** 89–107

BARTELINK D L 1931–2 Einrichtung zur Bildherstellung mittelst Röntgenstrahlen *Swiss Patent* 155930

BARTELINK D L 1932a Method of obtaining clear pictures by blotting out superfluous parts *Ned. Tijdschr. Geneesk.* **76** 23

BARTELINK D L 1932b Nouveau procédé radiographique de mise en évidence d'une région osseuse déterminée *J. Belge Radiol.* 21 447–50 (author referred to as Barteluck in paper)

BARTELINK D L 1933 Röntgenschnitte *Fortschr. Geb. Röntgenstr.* **47** 399–407

BATES R H T and Peters T M 1971 Towards improvements in tomography *NZ J. Sci.* **14** 883–96

BAYLIN G J 1944 Body section röntgenography *South. Med. J.* **37** 418–24

BELOT J 1937 Les nouveautés et les faits saillants présentés pour l'étude de la radiographie pulmonaire: à la 2nd Reunion Amicale Internationale *J. Radiol. Électrol.* **21** 16–20

BENDER R, Bellman S H and Gordon R 1970 ART and the ribosome: a preliminary report on the three-dimensional structure of individual ribosomes determined by an algebraic reconstruction technique *J. Theor. Biol.* **29** 483–7

BERRETT A, Brunner S and Valvassori G E 1973 *Modern Thin-section Tomography* (Springfield, Illinois: Charles C Thomas)

BERRY M V and Gibbs D F 1970 The interpretation of optical projections *Proc. R. Soc.* A **314** 143–52

BISTOLFI S 1934 Introduzione allo studio della stratigrafia *Radiol. Fis. Med.* **IA4** 439–69

BLASZCZYK M 1988 private communication (translation transcript, July)

BLEICH A R 1960 *The Story of X-rays from Röntgen to Isotopes* (New York: Dover) p 11

BLUM A 1971–5 Radiation detection device and a radiation detection method *US Patent* 3878373

BMJ 1937 Tomography *Br. Med. J.* 27 November 1075–6

BOAG J W 1984 Silvanus Phillips Thompson—some studies in the 'prehistory' of X-rays: The Silvanus Thompson Memorial Lecture, April 1983 *Br. J. Radiol.* **57** 1–15

BOAG J W and SMITHERS D W 1967 Foreword *Radioactive Isotopes in the Localisation of Tumours, Int. Nuclear Medicine Symp., London, September 1967* ed V R McCready, D M Taylor N G Trott, C B Cameron, E O Field, R J French and R P Parker (London: Heinemann)

BOCAGE A E M 1921–2 Procédé et dispositifs de radiographie sur plaque en mouvement *Fr. Patent 536464* (Engl. transl. Bricker J D 1964 *Classic Descriptions in Diagnostic Röntgenology* ed A J Bruwer (Springfield, Illinois: Charles C Thomas) pp 1414–17)

BOCAGE A E M 1938 Presentation d'appareil: le Biotome *Bull. Mém. Soc. Électrol. Radiol. Méd Fr.* **26** 210–6

BONTE G, TRINEZ G and BRENOT M 1955 *La Tomographie Axiale Transversale* (Paris: G Doin)

BOTH E 1936–9 Röntgengerät zur Abbildung von Körperschnitten *German Patent 686022*

BOWLEY A R, TAYLOR C G, CAUSER D A, BARBER D C, KEYES W I, UNDRILL P E, CORFIELD J R and MALLARD J R 1973 A radioisotope scanner for rectilinear, arc, transverse section and longitudinal section scanning: ASS—the Aberdeen Section Scanner *Br. J. Radiol.* **46** 262–71

BOYD D P 1974–6 Method and apparatus for X-ray or gamma ray 3D tomography using a fan beam *US Patent 3983398*

BOYD D P 1976–8 Position sensitive X-ray or gamma ray detector and 3D tomography using same *US Patent 4075491*

BOYD D P 1978 Position sensitive X-ray or gamma ray detector and 3D tomography using same *Austr. Patent AUS504935B*

BOZZETTI G 1935 La realizzazione pratica della stratigrafia *Radiol. Med.* **22** 257–67

BOZZO S R, ROBERTSON J S and MILAZZO J P 1968 A data processing method for a multidetector positron scanner *Fundamental Problems in Scanning* ed A Gottschalk and R N Beck (Springfield, Illinois: Charles C Thomas) pp 212–25

BRACEWELL R N 1956 Strip integration in radio-astronomy *Aust. J. Phys.* **9** 198–217

BRACEWELL R N and RIDDLE A C 1967 Inversion of fan-beam scans in radio-astronomy *Astrophys. J.* **150** 427–34

BRECHER R and BRECHER E 1969 *The Rays; A History of Radiology in the United States and Canada* (Baltimore, Maryland: The Williams and Wilkins Company) pp 258–63 (also quoting private communication from J R Andrews and from Mrs J Kieffer), pp 62–3

BREWSTER SIR D 1856 *Stereoscopy: Its History, Theory and Construction* (London: John Murray)

BRICKER J D 1964 Tomography *Classic Descriptions in Diagnostic Röntgenology* ed A J Bruwer 1964 (Springfield, Illinois: Charles C Thomas) pp 1406–12

BRIDGEMAN C F 1965 Use of radiation in philately and in examination of paintings *The Science of Ionising Radiation* ed E Etter (Springfield, Illinois:

Charles C Thomas) pp 655–81

BRILL A B 1989 private communication (letter to S Webb, 6 February)

BRILL A B, PATTON J A, ERICKSON J J and KING P H 1970–1 Multicrystal tomographic scanner for mapping thin cross sections of radioactivity in an organ of the human body *US Patent 3591806*

BROCK E H 1945 An adjustable and universally adaptable planigraph *Am. J. Röntgenol.* **54** 190–2

BROOKS R A and DI CHIRO G 1976 Principles of computer assisted tomography (CAT) in radiographic and radioisotopic imaging *Phys. Med. Biol.* **21** 689–732

BROWNELL G L 1988 private communication (letters to S Webb, 19 July, 16 November)

BROWNELL G L 1989 private communication (letter to S Webb, 17 January)

BROWNELL G L, ARONOW S and HINE G J 1966 Radioisotope scanning *Instrumentation in Nuclear Medicine* ed G J Hine (New York: Academic) pp 381–428

BROWNELL G L and BURNHAM C A 1974a MGH positron camera *Tomographic Imaging in Nuclear Medicine* ed G S Freedman (New York: Society of Nuclear Medicine) pp 154–64

BROWNELL G L and BURNHAM C A 1974b Recent developments in positron scintigraphy *Instrumentation in Nuclear Medicine* vol 2, ed G J Hine and J A Sorenson (New York: Academic) pp 135–59

BROWNELL G L, BURNHAM C A, HOOP B JR and KAZEMI H 1972 Positron scintigraphy with short-lived cyclotron-produced radiopharmaceuticals and a multicrystal positron camera *Medical Radioisotope Scintigraphy, Proc. Symp., Monte Carlo, October 1972* (Vienna: International Atomic Energy Agency) pp 313–30

BROWNELL G L, BURNHAM C A, WILENSKY S, ARONOW S, KAZEMI H and STREIDER D 1969 New developments in positron scintigraphy and the application of cyclotron-produced positron emitters *Medical Radioisotope Scintigraphy* (Vienna: International Atomic Energy Association) pp 163–76

BROWNELL G L and SWEET W H 1953 Localisation of brain tumours with positron emitters *Nucleonics* **11** 40–5

BRUCER M 1978 Nuclear medicine begins with a boa constrictor *J. Nucl. Med.* **19** 581–98

BRUNNETT C J 1975–7 Method and apparatus for improved radiation detection in radiation scanning systems *US Patent 4052620*

BRUWER A J 1964 *Classic Descriptions in Diagnostic Röntgenology* (Springfield, Illinois: Charles C Thomas) vols 1 and 2

BUDINGER T F 1974 Quantitative nuclear medicine imaging: applications of computers to the gamma camera and the whole body scanner *Progress in Atomic Physics* vol 4 *Recent Advances in Nuclear Medicine* ed J H Lawrence (New York: Grune and Stratton) pp 41–130

BUDINGER T F (ed) 1987 Fifty years of progress 1937–1987 *Donner Laboratory, Division of Biology and Medicine of the Lawrence Berkeley Laboratory, University of California, Berkeley, California, Publication* 268

BUDINGER T F, CROWE K M, CAHOON J L, ELISCHER V P, HUESMAN R H and KANSTEIN L L 1975 Transverse section imaging with heavy charged particles: theory and applications *Image Processing for 2D and 3D Reconstruction from Projections* (Stanford, California: Stanford University Press) pp MA1-1–MA1-4

BUFFÉ P 1936 Présentation de clichés de coupes radiographiques *Bull. Mém. Soc. Electrol. Radiol. Méd. Fr.* **25** 83–6

BULL J 1981 History of computed tomography *Radiology of the Skull and Brain: Technical Aspects of Computed Tomography* ed T H Newton and D G Potts (St Louis: C V Mosby Co) pp 3835–49

BURDON R S 1919 Binocular vision and radiography *Arch. Radiol. Electrotherapy* **24** 101–12

BURNHAM C A, BRADSHAW J F, KAUFMAN D E, CHESLER D A and BROWNELL G L 1982–5 Positron source position sensing detector and electronics *US Patent* 4531058

BURROWS E H 1986 *Pioneers and Early Years; a History of British Radiology* (Alderney: Colophon Press)

BUSH G B 1938 Some experiments in tomography *Br. J. Radiol.* **11** 611–22

BUSH G B 1939 Seriescopy: its principles and applications *Br. J. Radiol.* **12** 611–8

BUZZI G 1950 La stratigraphie axiale transversale dans la pathologie du médiastin *J. Radiol.* **31** 146–53

CAHILL M E 1941 Rectilinear body-section radiography *X-ray Technician* **13** 16–7

CASE J T 1912 The stereo röntgenography of the stomach and intestine *Arch. Röntgen Ray* **17** 46–9

CASE J T 1916 Stereoröntgenography of the alimentary tract *The Stereoclinic* ed H A Kelly (Troy, New York: Southworth Co) §34 (reviewed in 1916 *Arch. Radiol. Electrotherapy* **21** 99–100)

CASSEN B 1968 Problems of quantum utilisation *Fundamental Problems in Scanning* ed A Gottschalk and R N Beck (Springfield, Illinois: Charles C Thomas) pp 50–63

CASSEN B, CURTIS L, REED C W and LIBBY R 1951 Instrumentation for ^{131}I use in medical studies *Nucleonics* **9** 46–50

CHANG W, LIN S L, HENKIN R E and SALO B C 1980 A multisegmental slant-hole tomographic collimator (MUST): a new tomographic gamma camera system *J. Nucl. Med.* **21** P28

CHAOUL H 1935a Die röntgenologische Darstellbarkeit der einzelnen Lungenschlichten *Beit. Klin. Tuberculose Brauer* **86** 569–75

CHAOUL H 1935b Uber die Tomographie und insbesondere ihre

Anwendung in der Lungendiagnostik *Fortschr. Geb. Röntgenstr.* **51** 342–56

CHAOUL H 1935c Eine neues Röntgenuntersuchun gesverfahren zur Darstellung von Körperschlichten und seine Anwendung in der Lungendiagnostik (Tomographie) *Fortschr. Geb. Röntgenstr.* **52** 28–43

CHAOUL H and GREINEDER K 1936 Lungenkarzinom und Lungenabzes im tomographischen Bild *Fortschr. Geb. Röntgenstr.* **53** 232–9

CHARKES N D and WILLIAMS J L 1974 Stereoscintiphotography: Application in brain scanning and experimental studies in depth perception *Tomographic Imaging in Nuclear Medicine* ed G S Freedman (New York: Society of Nuclear Medicine) pp 186–95

CHAUSSE C 1939 Applications of the stereoradiographic centring apparatus in otology and general surgery *Br. J. Radiol.* **12** 76–90

CLARKE K C 1939 *Positioning in Radiography* (London: Heinemann)

COLYER C 1937 Tomography in the vertical position *Lancet* **2** 1302–3

CORMACK A M 1963 Representation of a function by its line integrals, with some radiological applications *J. Appl. Phys.* **34** 2722–7

CORMACK A M 1964 Representation of a function by its line integrals, with some radiological applications 2 *J. Appl. Phys.* **35** 2908–13

CORMACK A M 1980 Early two dimensional reconstruction and recent topics stemming from it (Nobel Prize lecture) *Science* **209** 1482–6

CORMACK A M and KOEHLER A M 1976 Quantitative proton tomography: preliminary experiments *Phys. Med. Biol.* **21** 560–9

COTTENOT P 1938 Thoracic serioscopy *Radiology* **31** 1–7

COTTRALL M F and FLIONI-VYZA A 1972 Design of a scintillation camera tomographic system and investigations of its performance *Medical Radioisotope Scintigraphy, Proc. Symp., Monte Carlo, October 1972* (Vienna: International Atomic Energy Agency) pp 381–408

COULAM C M, ERICKSON J J and GIBBS S J 1981 Image and equipment considerations in conventional tomography *The Physical Basis of Medical Imaging* ed C M Coulam, J J Erickson, F D Rollo and A E James Jr (Norwalk, Connecticut: Appleton Century Crofts) pp 123–40

COX J R and SNYDER D L 1975–6 Tomography system having axial scanning *US Patent* 3983399

CRAMER H and WOLD H 1936 Some theorems on distribution functions *J. London Math. Soc.* **11** 290–4

CROFT B Y 1986 *Single Photon Emission Computed Tomography* (Chicago, Illinois: Year Book Medical Publishers)

CROWE K M, BUDINGER T F, CAHOON J L, ELISCHER V P, HUESMAN R H and KANSTEIN L L 1975 Axial scanning with 900 MeV alpha particles *IEEE Trans. Nucl. Sci.* **NS-22** 1752–4

CROWTHER R A, DE ROSIER D J and KLUG A 1970 The reconstruction of a three-dimensional structure from projections and its application to electron microscopy *Proc. R. Soc.* A **317** 319–40

CZERMAK P 1896 cited in Stereoscopic pictures with the Röntgen rays *Photography* **8** 836–7; cited in Keats T E 1964 Origins of stereoscopy and diagnostic röntgenology *Classic Descriptions in Radiology* ed A J Bruwer (Springfield, Illinois: Charles C Thomas) pp 983–6

DE ABREU M 1926 *Essai sur une Nouvelle Radiologie Vasculaire* (Paris: Masson)

DE ABREU M 1930 *Etudes Radiologiques sur le Poumon et le Mediastin* (Paris: Masson)

DE ABREU M 1944 Enfisema orbitario *Radiologica* **7** 221–5

DE ABREU M 1948 Theory and technique of simultaneous tomography *Am. J. Röntgenol.* **60** 668–74

DE ROSIER D J and KLUG A 1968 Reconstruction of three-dimensional structures from electron micrographs *Nature* **217** 130–4

DE VULPIAN P 1951 Considérations sur la tomographie *J. Radiol., Électrol., Méd. Nucl.* **32** 376–9

DEWING S B 1962 *Modern Radiology in Historical Perspective* (Springfield, Illinois: Charles C Thomas) pp 98–100

DI CHIRO G 1963 Axial transverse encephalography *US Department of Health, Education and Welfare Public Health Service Publication* 1104

DI CHIRO G, JOHNSTON G S, MIRALDI F D and YON E T 1974 The Tomoscanner: clinical experience *Tomographic Imaging in Nuclear Medicine* ed G S Freedman (New York: Society of Nuclear Medicine) pp 250–5

DIDIEE J 1944 Position actuelle de la tomographie en radiodiagnostic *Bull. Soc. Fr. Électrothérapie Radiol., J. Radiol.* **26** 328–32

DISTLER W, KINTOPP E and LINKE G 1975–6 Patients' support installation for a tomographic X-ray apparatus *US Patent* 3974388

DRUMMOND D H and SCHMELA W W 1939 The computation of dimensions in planigraphy with mathematical instruments *Radiology* **32** 550–5

DUNHAM K 1916 *Stereoröntgenography in Pulmonary Diagnosis* (Troy, New York: The Southworth Co.) (reviewed in *Arch. Radiol. Electrotherapy* **21** 272–4)

EDHOLM P 1960 The tomogram; its formation and content *Acta Radiol., Suppl.* **193** 1–109

EDHOLM P, GÖSTA HELLSTROM L and JACOBSON B 1978 Tomography with optical reconstruction *Phys. Med. Biol.* **23** 90–9

ELL P J 1982 The brain. The role of SPECT in the diagnosis of space-occupying disease *Computed Emission Tomography* ed P J Ell and B L Holman (Oxford: OUP) pp 399–418

EMI 1972 EMI Limited—The MacRobert Award 1972; computerized transverse axial tomography *Award Citation Prepared by EMI Limited*

ETTER E 1965 Historical data relating to discovery of X-rays *The Science of Ionising Radiation* ed E Etter (Springfield, Illinois: Charles C Thomas) pp 18–38 (quoting Sylvanus Thompson at the Röntgen Society of London, 5 November 1897)

EVANS N T S, KEYES W I, SMITH D, COLEMAN J, CUMPSTEY D, UNDRILL

P E, ETTINGER K V, ROSS K, NORTON M Y, BOLTON M P, SMITH F W and MALLARD J R 1986 The Aberdeen Mark 2 single photon emission tomographic scanner: specification and some clinical applications *Phys. Med. Biol.* **31** 65–78

FARR R F, SCOTT A C H, OLLERENSHAW R and EVERARD G J H 1964 *Transverse Axial Tomography* (Oxford: Blackwell)

FEINDEL W and YAMAMOTO Y L 1978 Physiological tomography by positrons: introduction and historical note *J. Comput.-Assisted Tomogr.* **2** 637–8

FELSON B 1977 A man of style: E R N Grigg obituary *Semin. Röntgenol* **12** 1–2

FERGUSON J W 1936 Röntgen stereoscopy *Am. J. Röntgenol.* **35** 662–9

FISCHGOLD H and BULL J W D 1967 A short history of neuroradiology *Proc. 8th Symp. Neuroradiologicum* (Schering)

FRAIN and LACROIX 1947a Courbe-enveloppe et coupes horizontales *Bull. Soc. Fr. Électroradiol. Méd., J. Radiol.* **28** 142–3

FRAIN and LACROIX 1974b Etude experimentale sur l'obtention de coupes horizontales *Paris Med.* **37** 94–5

FRAIN and LACROIX 1947c Effet stratigraphique et coupes horizontales *Presse Méd.* **18** 205

FRAIN and LACROIX 1948 De l'obtention de coupes horizontales *J. Radiol. Electrol.* **29** 256–7

FRANK G 1940 Verfahren zur Herstellung von Körperscnittbildern mittels Röntgenstrahlen *German Patent* 693374

FRANKE K 1977–9 Tomographic apparatus for producing transverse layer images *US Patent* 4150293

FREEDMAN G S 1970 Tomography with a gamma camera *J. Nucl. Med.* **11** 602–4

FREEDMAN G S 1972 Gamma camera tomography—preliminary clinical experience *Radiology* **102** 365–9

FREEDMAN G S 1974 Digital gamma camera tomography—theory *Tomographic Imaging in Nuclear Medicine* ed G S Freedman (New York: Society of Nuclear Medicine) pp 68–75

FROGGATT R J 1976–8 Treatment of absorption errors in computerized tomography *US Patent* 4081681

FROGGATT R J 1977–8 Radiographic apparatus *US Patent* 4118631

FROGGATT R J and PERCIVAL W S 1975–6 Data acquisition in tomography *US Patent* 3996467

GASSUL R 1936 Anatomia *in vivo* Tomografia *Sovietskii Vrachebnyi Zhurnal* **21** 1618–26

GEBAUER A 1949 Körperschichtaufnahmen in transversalen (horizontalen) Ebenen *Fortschr. Geb. Röntgenstr.* **71** 669–96

GEBAUER A 1956 Historical development of radiological tomography *Indian J. Radiol.* (souvenir number) 213–27

GIACOBINI E and MANZI G 1954 Su di un nuovo apparecchio universale per la stratigrafia in posizione orizzontale, obliqua, verticale e per assiale trasversa con paziente fermo *Radiol. Med.* **40** 37−53

GILBERT P F C 1972a Iterative methods for the three dimensional reconstruction of an object from projections *J. Theor. Biol.* **36** 105−17

GILBERT P F C 1972b The reconstruction of a three-dimensional structure from projections and its application to electron microscopy 2: Direct Methods *Proc. R. Soc.* B **182** 89−102

GILLESPIE J E and ISHERWOOD I 1986 Three dimensional anatomical images from computed tomographic scans *Br. J. Radiol.* **59** 289−92

GLASSER O 1933 *The Science of Radiology* (London: Ballière, Tindall and Cox)

GLASSER O 1938 The life of Wilhelm Conrad Röntgen as revealed in his letters *1938 Yearbook of Radiology* ed C A Waters and I I Kaplan (Chicago, Illinois: Year Book Publishers) pp 9−17

GLASSER O 1965 Wilhelm Conrad Röntgen and the discovery of the Röntgen rays *The Science of Ionising Radiation* ed E Etter (Springfield, Illinois: Charles C Thomas) pp 5−17

GMITRO A F, GREIVENKAMP J E, SWINDELL W, BARRETT H H, CHIU M Y and GORDON S K 1980 Optical computers for reconstructing objects from their X-ray projections *Opt. Eng.* **19** 260−72

GOITEIN M 1972 Three-dimensional density reconstruction from a series of two-dimensional projections *Nucl. Instrum. Methods* **101** 509−18

GOITEIN M 1989 private communication (discussion 5 July 1989)

GOLDBERG B B and KIMMELMAN B A 1988 *Medical Diagnostic Ultrasound: a Retrospective on its 40th Anniversary* (Rochester, New York: Eastman−Kodak Company) pp 1−49

GOLDMAN C H 1943 A tomographic attachment suitable for most X-ray plants *Br. J. Radiol.* **16** 355−6

GREINEDER K 1935 Die Tomographie der normalen Lunge *Fortschr. Geb. Röntgenstr.* **52** 443−61

GRIESBACH R and KEMPER F 1955 *Röntgen Schlichtverfahren; grundlagen der technischen entwicklung und der Klinischen Anwendung für die Praxis* (Stuttgart: Georg Thieme)

GRIGG E R N 1965 *The Trail of the Invisible Light; from X-Strahlen to Radio(bio)logy* (Springfield, Illinois: Charles C Thomas)

GROSSMANN G 1934 Procédé et dispositif pour la représentation radiographique des sections des corps *Fr. Patent* 771887

GROSSMANN G 1935a Lung tomography *Br. J. Radiol.* **8** 733−51

GROSSMANN G 1935b Tomographie 1; Röntgenographische Darstellung von Körperschnitten *Fortschr. Geb. Röntgenstr.* **51** 61−80

GROSSMANN G 1935c Tomographie 2: Theoretisches über Tomographie *Fortschr. Geb. Röntgenstr.* **51** 191−208

GROSSMANN G 1935d Bemerkungen zu vorstehendem Aufsatz von Karol Mayer 'Zur Tomographie' *Fortschr. Geb. Röntgenstr.* **52** 624

GROSSMANN G 1935e Praktische Voraussetzungen für die Tomographie *Fortschr. Geb. Röntgenstr.* **52** 44

GROSSMANN G 1936–8 Apparatus for making radiographs *US Patent* 2110954

GROVER H W 1973 Reflections on early X-ray engineering *Br. J. Radiol.* **46** 757–61

HAENISCHE G F 1911 Röntgenological impressions of a journey in the United States *Arch. Röntgen Ray* **15** 328–36

HANDMAKER H, ANGER H O and MCRAE J 1974 Rotational cinescinti-photography as a means of obtaining three dimensional organ images *Tomographic Imaging in Nuclear Medicine* ed G S Freedman (New York: Society of Nuclear Medicine) pp 196–9

HARPER P V, BECK R N, CHARLESTON D E, BRUNSDEN B and LATHROP K A 1965 The three dimensional mapping and display of radioisotope distributions *J. Nucl. Med.* **6** 332

HARPER P V 1968 The three dimensional reconstruction of isotope distributions *Fundamental Problems in Scanning* ed A Gottschalk and R N Beck (Springfield, Illinois: Charles C Thomas) pp 191–4

HART W E 1965 Focusing collimator coincidence scanning *Radiology* **84** 126

HASSELWANDER A 1954 *Die Objective Stereoskopie an Röntgenbildern* (Stuttgart: George Thieme)

HECKMANN K 1939 Die Röntgenperspektive und ihre Umwandlung durch eine neue Aufnahmetechnik *Fortschr. Geb. Röntgenstr.* **60** 144–57

HERDNER R 1948a Révélations tomographiques sur la base du crâne. Premières coupes horizontales réalisées sur le vivant. Leur intérêt chirurgical. *Mém. Acad. Chir.* **5** 115–23

HERDNER R 1948b Exploration tomographique du sternum de l'adulte et des articulations sterno-claviculaires. Trois cas de tuberculose ostéo-articulaire du sternum *Mém Acad. Chir.* **5** 40

HERMAN G T 1972 Two direct methods for reconstructing pictures from their projections: a comparative study *Comput. Graph. Image Process.* **1** 123–44

HERMAN G T 1980 *Image Reconstruction from Projections: The Fundamentals of Computed Tomography* (New York: Academic)

HILL E G and BARNARD T W 1919 On the distortion of stereoscopic images *Arch. Radiol. Electrotherapy* **24** 112–6

HILLEMAND P 1953 André Bocage (1892–1953) obituary *Presse Méd.* **61** 1496

HODGES P C 1945 Development of diagnostic X-ray apparatus during the first fifty years *Radiology* **45** 438–48

HOLLAND C T 1916 Review of book by Davidson Sir J M 1916 *Localisation by X-rays and stereoscopy* H K Lewis and Co Ltd, London *Arch. Radiol. Electrotherapy* **21** 98–9

HOLLAND C T 1919 Notes on a special tube stand for rapid X-ray stereoscopic work constructed for the Cancer Hospital *Arch. Radiol. Electrotherapy* **24** 218–26

HOLT J F, HODGES F J, JACOX H W and KLIGERMAN M M 1955–6 Use of pantomography in clinical investigations *1955–6 Yearbook of Radiology* (Chicago, Illinois: Year Book Publishers) pp 29–30

HOLLY E W 1942 Some radiographic technics in which movement is utilised *Radiogr. Clin. Photogr.* **18** 78–83

HOPF M 1943 Eine neue Apparatur für Stereo-Röntgenoskopie *Radiol. Clin.* **12** 65–9

HOUNSFIELD G N 1968–72 A method of and apparatus for examination of a body by radiation such as X or gamma radiation *UK Patent* 1283915

HOUNSFIELD G N 1971–3 Method and apparatus for measuring X or gamma radiation absorption or transmission at plural angles and analyzing the data *US Patent* 3778614

HOUNSFIELD G N 1973 Computerised transverse axial scanning (tomography): Part 1 Description of system *Br. J. Radiol.* **46** 1016–22

HOUNSFIELD G N 1973–5a Penetrating radiation examining apparatus in combination with body locating structure *US Patent* 3881110

HOUNSFIELD G N 1973–5b Penetrating radiation examining apparatus having a scanning collimator *US Patent* 3866047

HOUNSFIELD G N 1973–7 Improvements in or relating to radiography *UK Patent* 1468810

HOUNSFIELD G N 1974–5a Method of and apparatus for examining a body by radiation such as X or gamma radiation *US Patent* 3919552

HOUNSFIELD G N 1974–5b Method of and apparatus for examining a body by radiation such as X or gamma radiation *US Patent* 3924131

HOUNSFIELD G N 1974–6a Apparatus for examining a body by radiation such as X or gamma radiation *US Patent* 3944833

HOUNSFIELD G N 1974–6b Radiography *US Patent* 3934142

HOUNSFIELD G N 1974–6c Apparatus for examining objects by means of penetrating radiation *US Patent* 3940625

HOUNSFIELD G N 1974–6d Apparatus for examining a body by means of penetrating radiation *US Patent* 3999073

HOUNSFIELD G N 1974–77 Radiography *US Patent* 4035647

HOUNSFIELD G N 1975–77 Data acquisition in tomography *US Patent* 4002911

HOUNSFIELD G N 1980 Computed medical imaging (Nobel Prize lecture) *Science* **210** 22–8

HOUNSFIELD G N 1989 private communication (letter to S Webb, 10 March)

HUESMAN R H, GULLBERG G T, GREENBERG W L and BUDINGER T F 1977 RECLBL Library users manual; Donner algorithms for reconstruction tomography *Lawrence Berkeley Laboratory Publication* 214

IMBERT A and BERTIN-SANS H 1896 Stereoscopic photographs obtained by means of X-rays (extract) *C. R. Acad. Sci., Paris* 30 March 786

INOUYE T and MIZUTANI H 1978–80 Method and apparatus for performing computed tomography *US Patent* 4205375

JACHES L, STEWART W H and IMBODEN H M 1916 *The American Atlas of Stereoröntgenography* (Troy, New York: The Southworth Co) (reviewed in *Arch. Radiol. Electrotherapy* **21** 164–5)

JACOBS J E, CORNERS H and HOWELL J F 1955–9 Translation system *US Patent* 2907883

JACOBS L G 1938 Röntgenography of second cervical vertebra by Otonello's method *Radiology* **31** 412–3

JAMA 1965 André Bocage (1892–1953)—French tomographer, editorial, *J. Am. Med. Assoc.* **193** 233

JANKER R 1950 Ein Universal-Schichtaufnahmegerat *Fortschr. Geb. Röntgenstr.* **73** 253–61

JARRE H A and TESCHENDORF O E W 1933 Röntgen-stereoscopy; a review of its present status *Radiology* **21** 139–55

JASZCZAK R J, MURPHY P H, HUARD D and BURDINE J A 1977 Radionuclide emission computed tomography of the head with 99m Tc and a scintillation camera *J. Nucl. Med.* **18** 373–80

JCAT 1978 *Proc. 1st Int. Symp. on Positron Emission Tomography, Montreal, Quebec, 2–3 June, 1978, J. Comput.-Assisted Tomogr.* **2** 637–64

JOHNSON C R 1935 Röntgen mensuration by stereoröntgenometry *Radiology* **25** 492–4

JONES W G and BRADLEY-BOWRON E W 1939 A home-made 'Tomoscope' *Indian Med. Gaz.* **74** 618–21

JÖNSSON G 1937 A method of obtaining structural pictures of the sternum *Acta Radiol.* **18** 336–40

JUPE M H 1946 The physicist in the radiodiagnostic department *Br. J. Radiol.* **119** 301–3

KALENDER W, LINKE G and PFEILER M 1980–2 Tomographic X-ray apparatus for the production of transverse layer images *US Patent* 4324978

KAMEN M D 1957 The isotopes of oxygen, nitrogen, phosphorus, and sulfur *Isotopic Tracers in Biology: An Introduction to Tracer Methodology* (Orlando, Florida: Academic) p 339

KAUFMAN J 1936a Planeography, localisation and mensuration: 'standard depth curves' *Radiology* **27** 168–74

KAUFMAN J 1936b The planeogram *Radiology* **27** 732–5

KAUFMAN J 1938 Object reconstruction by planigraphy: reconstruction and localisation of planes *Radiology* **30** 763–5

KAUFMAN J and KOSTER H 1940 Serial planeography (serioscopy) and serial planigraphy *Radiology* **34** 626–9

KEATS T E 1964 Origins of stereoscopy and diagnostic röntgenology *Classic Descriptions in Diagnostic Röntgenology* ed A J Bruwer (Springfield, Illinois: Charles C Thomas) pp 983–6

KENNY P J 1971 Spatial resolution and count rate capacity of a positron camera: some experimental and theoretical considerations *Int. J. Appl. Radiat. Isot.* **22** 21–8

KENNY P J, MYERS M J, LUNDY A, RITTER F and LAUGHLIN J S 1968 Digital Anger camera with electronic focusing for positron emitters *Fundamental Problems in Scanning* ed A Gottschalk and R N Beck (Springfield, Illinois: Charles C Thomas) pp 226–8

KERLEY P 1937 X ray diagnosis *Practitioner* **139** 483–7

KEYES J W 1982 Instrumentation *Computed Emission Tomography* ed P J Ell and B L Holman (Oxford: OUP) pp 243–62

KEYES J W 1988 private communication (letter to S Webb, 12 July)

KEYES J W, ORLANDEA N, HĖETDERKS W J, LEONARD P F and ROGERS W L 1977 The Humongotron—a scintillation camera transaxial tomograph *J. Nucl. Med.* **18** 381–7

KIEFFER J 1929–34 X-ray device and method of technique *US Patent* 1954321

KIEFFER J 1938 The laminagraph and its variations: Applications and implications of the planigraphic principles *Am. J. Röntgenol.* **39** 497–513

KIEFFER J 1939a Body section radiographic technic *The X-ray Technician* July 12–16

KIEFFER J 1939b Analysis of laminographic motions and their values *Radiology* **33** 560–85

KIEFFER J 1943 The general principles of body section radiography *Radiogr. Clin. Photogr.* **19** 2–10

KIMBLE H E 1935 A simplified mechanical method for radiographic mensuration and localisation *Radiology* **24** 39–46

KING P H, PATTON J A, PICKENS D R, SWEENEY J and BRILL A B 1971 A multidetector orthogonally coincident focal point tomographic scanner *South. Med. J.* **64** 1422

KISHI K, KURIHARA S, YUASA M and GOUKE H 1969 Radiotherapy treatment planning apparatus *Toshiba Rev.* **43** 36–41

KORENBLYUM B I, TETEL'BAUM S I, TYUTIN A A 1958 About one scheme of tomography *Izv. Vyssh. Uchebn. Zaved., Radiofiz.* **1** 151–7 (translated from the Russian by Professor H H Barrett, Tucson, Arizona)

KUHL D E 1968 The current status of tomographic scanning *Fundamental Problems in Scanning* ed A Gottschalk and R N Beck (Springfield, Illinois: Charles C Thomas) pp 179–90

KUHL D E 1989 Concerning the role of Dr David Kuhl MD in the evolution of tomographic imaging of ionizing radiation, private communication (letter to S Webb, 27 April)

KUHL D E and EDWARDS R Q 1963 Image separation radioisotope scanning *Radiology* **80** 653–61

KUHL D E and EDWARDS R Q 1969 Digital processing for modifying and rearranging rectilinear and section scan data under direct observation *Medical Radioisotope Scintigraphy* (Vienna: International Atomic Energy Agency) pp 703–13

KUHL D E and EDWARDS R Q 1970 The Mark 3 scanner: a compact device for multiple view and section scanning of the brain *Radiology* **96** 563–70

KUHL D E, EDWARDS R Q and RICCI A R 1974 Transverse section scanner at the University of Pennsylvania *Tomographic Imaging in Nuclear Medicine* ed G S Freedman (New York: Society of Nuclear Medicine) pp 19–27

KUHL D E, EDWARDS R Q, RICCI A R and REIVICH M 1972 Quantitative section scanning *Medical Radioisotope Scintigraphy, Proc. Symp., Monte Carlo, October 1972* (Vienna: International Atomic Energy Agency) pp 347–53

KUHL D E, HALE J and EATON W L 1966 Transmission scanning: a useful adjunct to conventional emission scanning for accurately keying isotope deposition to radiographic anatomy *Radiology* **87**, 278–84

LANGLEY J R 1944 *Medical history, Department of radiology, 6th General Hospital, 24 February–15 June 1944* (in Allen *et al* 1966)

LASSER E C and NOWAK E L 1956 Multiple simultaneous body-section radiography *Radiology* **66** 577–81

LAUGHLIN J S 1983 Historical review article: AAPM and RAMPS— antecedents and perspectives *Med. Phys.* **10** 387–94

LE CLERK J L, HARVEY H S, LANDA L S and RYKEL J E 1967–70 Test object and cassette for tomography *US Patent 3509337*

LEMAY C A G 1975–8 Radiography apparatus with photocell drift compensating means *US Patent 4071760*

LEMAY C A G 1976–7 Radiography *US Patent 4031395*

LEVY L M and OKEZIE O 1963 Three dimensional scintiscanning *J. Nucl. Med.* **4** 181

LIEBETRUTH R 1977–9 Tomographic X-ray apparatus for the production of transverse layer images *US Patent 4174481*

LILL B H 1977–8 Apparatus for computerized tomography having improved antiscatter collimators *US Patent 4101768*

LINDEGAARD-ANDERSEN A and THUESON G 1978 Transaxial tomography camera with built in photographic image reconstruction *J. Phys. E. Sci. Instrum.* **11** 805–11

LITTLETON J T 1973 Tomographic equipment *Modern Thin Section Tomography* ed A Berrett, S Brunner and G E Valvassori (Springfield, Illinois: Charles C Thomas) pp 28–47

LITTLETON J T 1976 *Tomography: Physical Principles and Clinical Applications* Section 17 of *Golden's Diagnostic Radiology* ed L L Robbins (Baltimore, Maryland: The Williams and Wilkins Company)

LITTLETON J T and RAVENTOS A 1965 Equipment of all types *The Science of Ionising Radiation* ed E Etter (Springfield, Illinois: Charles C Thomas) pp 64–100

LITTLETON J T, RUMBAUGH C L and WINTER E S 1963 Polydirectional body section röntgenography *Am. J. Röntgenol.* **89** 1179–93

LUX P 1978–80 Device for computed tomography *US Patent* 4241404

MACKAY A L 1977 *The Harvest of a Quiet Eye* (Bristol: Adam Hilger) (quoting Sir William Osler (1849–1919) in *Books and Man* and George Orwell in *Nineteen Eighty Four*)

MACKAY R S 1984 Computed tomography *Medical Images and Displays. Comparisons of Nuclear Magnetic Resonance, Ultrasound, X-ray and Other Modalities* (New York: Wiley)

MACKENZIE DAVIDSON SIR J 1898 Remarks on the value of stereoscopic photography and skiagraphy; records of clinical and pathological appearance *Br. Med. J.* **2** 1668–71

MACKENZIE DAVIDSON SIR J 1914 (no title for article) *Arch. Radiol. Electrotherapy* **19** 1–2

MACKENZIE DAVIDSON SIR J 1916 *Localisation by X-rays and Stereoscopy* (London: H K Lewis)

MACKENZIE DAVIDSON SIR J 1919 Stereoscopic radiography (text of lecture 18 October 1918) *Arch. Radiol. Electrotherapy* **23** 340–6

MACKENZIE DAVIDSON SIR J 1919 Obituary *Arch. Radiol. Electrotherapy* **23** 337–40

MANGES W F 1911 Description of method of measuring female pelvis *Am. Quart. Röntgenology* April 41–3

MASSEY J 1988 private communication (letter to S Webb, 4 May)

MASSIOT M 1935 Essais d'un appareil pour radiographie analytique *Soc. Radiol. Méd. Paris* No 221 395–8

MASSIOT M J 1938 Présentation du biotome du dr Bocage et du planigraphie du docteur Ziedses des Plantes *Bull. Mém. Soc. Électrol. Radiol. Méd. Fr.* **26** 520–3

MASSIOT J 1936 Étude comparative des méthodes de radiographie analytique *Bull. Mém. Soc. Radiol. Méd. Fr.* **24** 394–402

MASSIOT J 1937 La sériescopie: méthode d'étude analytique par plans successifs *Bull. Mém. Soc. Radiol. Méd. Fr.* **25** 390–3

MASSIOT J 1974 Historique de la Tomographie. Plaquette by care of the Society Massiot Philips on occasion of the *Journées Nationales de Radiologie* 1–15 (in French) (Similar but not identical text in English in Massiot J 1974 *Med. Mundi* **19** 106–15) unpublished

MASSIOT-PHILIPS undated but circa 1950 Radiotome *Industrial Technical Publication* 3/6

MAYER K 1916 *Radyologiczne Rözpozanie Rozniczkowe Chorób Serca i Aorty* (Krakow: Gebethner)

MAYER K 1935a Zur Tomographie *Fortschr. Geb. Röntgenstr.* **52** 622–3

MAYER K 1935b Erwiderung auf die Bemerkungen von Dr G Grossmann *Fortschr. Geb. Röntgenstr.* **52** 624

MAYER K 1935c Erwiderung auf die Bemerkungen von A Vallebona und St Bistolfi, Genua *Fortschr. Geb. Röntgenstr.* **52** 625

MAYNEORD W V 1952 The radiography of the human body with radioactive isotopes *Br. J. Radiol.* **25** 517–25

MAYNEORD W V, TURNER R C, NEWBERY S P and HODT H J 1951 A method of making visible the distribution of activity in a source of ionising radiation *Nature* **168** 762–5

MAYO B J and BEST J E 1975–7 Tomography *US Patent* 4032761

MCAFEE J G, MOZLEY J M, NATARAJAN T K, FUEGER G F and WAGNER H N Jr 1966 Scintillation scanning with an eight inch diameter sodium iodide (Tl) crystal *J. Nucl. Med.* **7** 521–47

MCAFEE J G, MOZLEY J M and STABLER E P 1969 Longitudinal tomographic radioisotopic imaging with a scintillation camera: theoretical considerations of a new method *J. Nucl. Med.* **10** 654–9

MCAFEE J G, MOZLEY J M and STABLER E P 1970–3 Tomographic radioisotopic imaging with a scintillation camera *US Patent* 3714429

MCDOUGALL J B 1936 The tomograph; its use in pulmonary tuberculosis *Lancet* **25** 185–7

MCDOUGALL J B 1937 Discussion on the clinical value of the tomograph *Proc. R. Soc. Med.* **31** 379–96

MCDOUGALL J B 1940 *Tomography* (London: H K Lewis)

MCDOUGALL J B 1945 Tomography *Edinburgh Med. J.* **52** 127–31

MCDOUGALL J B and CRAWFORD J H 1938 Further experiences with tomography in pulmonary tuberculosis *Br. Med. J.* 15 October 782–3

MCGRIGOR D B 1939 Selected methods of foreign body localisation adaptable to radiology in wartime *Br. J. Radiol.* **12** 619–31

MCINNES 1954 Tomography *Radiography* **22** 43–52

MCRAE J and ANGER H O 1974 Tomographic imaging in nuclear medicine *Progress in Atomic Physics* vol 4 *Recent Advances in Nuclear Medicine* ed J H Lawrence (New York: Grune and Stratton) pp 189–213

MCWHIRTER R 1988 private communication (letter to S Webb, 15 November)

MERCER E G 1988 private communication (letters to S Webb, 14 April and 24 April)

MESERVEY A B 1938 Depth effects in röntgenograms *Am. J. Röntgenol.* **39** 439–49

MIRALDI F D 1971–4 Tomographic scanning process *US Patent* 3784820

MIRALDI F D and DI CHIRO G 1970 Tomographic techniques in radio-isotope imaging with a proposal of a new device: the Tomoscanner *Radiology* **94** 513–20

MIRALDI F D, DI CHIRO G and SKOFF G 1969 Evaluation of current methods of radioisotope tomography and design of a new device: the

Tomoscanner *J. Nucl. Med.* **10** 358–9

MIRALDI F D, YON E T and DI CHIRO G 1974 Tomoscanner: theory and prototype *Tomographic Imaging in Nuclear Medicine* ed G S Freedman (New York: Society of Nuclear Medicine) pp 44–56

MITCHELL W 1910 Correspondence: Stereoscopic Röntgenography *Arch. Röntgen Ray* **15** 243–4

MOLLER P F 1968 *History and Development of Radiology in Denmark 1896–1950* (Copenhagen: NYT Nordisk Forlag-Arnold Busck) pp 373, 440

MONAHAN W G and POWELL M D 1974 Three-dimensional imaging of radionuclide distributions by gamma–gamma coincidence detection *Tomographic Imaging in Nuclear Medicine* ed G S Freedman (New York: Society of Nuclear Medicine) pp 165–75

MOODY I 1970 *The Society of Radiographers: 50 years of History* (London: Society of Radiographers)

MOREL-KAHN and BERNARD J 1935 Une nouvelle technique radiographique: La représentation 'en coupe' de l'organisme *Press Méd.* **79** 1525–6

MOORE S 1938 Body section röntgenography with the laminograph *Am. J. Röntgenol.* **39** 514–22

MOORE S 1939 Body section radiography *Radiology* **33** 605–14

MUEHLLEHNER G 1970–2 Tomographic imaging device using a rotating slanted multichannel collimator *US Patent* 3684886

MUEHLLEHNER G 1971 A tomographic scintillation camera *Phys. Med. Biol.* **16** 87–96

MUEHLLEHNER G 1973–4 Tomographic imaging device *US Patent* 3852603

MUEHLLEHNER G 1974 Performance parameters for a tomographic scintillation camera *Tomographic Imaging in Nuclear Medicine* ed G S Freedman (New York: Society of Nuclear Medicine) pp 76–83

MUEHLLEHNER G 1988 private communication (letter to S Webb, 10 August)

MUEHLLEHNER G, ATKINS F and HARPER P V 1977 Positron camera with longitudinal and transverse tomographic capabilities *Proc. Symp. on Radionuclide Imaging, Los Angeles, California, 25–29 October 1976* vol 1 (Vienna: International Atomic Energy Agency) pp 291–307

MUEHLLEHNER G and WETZEL R A 1971 Section imaging by computer calculation *J. Nucl. Med.* **12** 76–84

MUIR J 1924 *A Manual of Practical X-ray Work* (London: Heinemann)

MYERS M J, KEYES W I and MALLARD J R 1972 An analysis of tomographic scanning systems *Medical Radioisotope Scintigraphy, Proc. Symp., Monte Carlo, October 1972* (Vienna: International Atomic Energy Agency) pp 331–45

NAUD A 1936 Contribution à l'étude d'une nouvelle méthode radiographique. Radiographies en coupes de l'organisme *Thèse,* Lyon, Faculté

de Médecine et de Pharmacie de Lyon. Année Scolaire 1936–7 No 78
NS 1972 X-ray diagnosis peers inside the brain *New Sci.* 27 April

OLDENDORF W H 1960–3 Radiant energy apparatus for investigating selected areas of the interior of objects obscured by dense material *US Patent* 3106640
OLDENDORF W H 1961 Isolated flying spot detection of radiodensity discontinuities; displaying the internal structural pattern of a complex object. *IRE Trans. Biomed. Electron.* **BME-8** 68–72
OLDENDORF W H 1980 *The Quest for an Image of Brain: Computerized Tomography in the Perspective of Past and Future Imaging Methods* (New York: Raven Press)
OLSSON O 1942 Eine neue methode zur isolierten oberflächenradiographie *Acta Radiol.* **23** 420–2
OLSSON O 1944 Eine methode zur radiographie oberflächlichter schlichten *Acta Radiol.* **25** 701–18
OTT P 1935 Die gegenwärtige Leistungsfähigkeit der Körperschichtdarstellungen *Fortschr. Geb. Röntgenstr.* **52** 40–3
OTT R J, FLOWER M A, BABICH J W and MARSDEN P K 1988 The physics of radioisotope imaging *The Physics of Medical Imaging* ed S Webb (Bristol: Adam Hilger) pp 142–318
OTONELLO P 1930 Nuovo metodo per la radiografia della colonna cervicale completa in proiezione sagittale ventro-dorsale *Riv. Radiol. Fis. Med.* **2** 291–4
OTONELLO P 1935 Tecnica radiografia ad ampulla mobile *Ann. Radiol.* **9** 22

PAATERO Y V 1949 A new tomographical method for radiographing curved outer surfaces *Acta Radiol.* **32** 177–84
PAATERO Y V 1950–4 Method of and apparatus for X-ray photographing curved surfaces, especially for medical purposes *US Patent* 2684446
PAATERO Y V 1955a Die Anwendung der Pantomographie für klinische Untersuchungen *Fortschr. Geb. Röntgenol* **82** 525–8
PAATERO Y V 1955b The principles of the construction and function of the stereo-pantomograph *Acta Radiol.* **43** 113–9
PATTON J A, BRILL A B, ERICKSON J J, COOK W E and JOHNSTON R E 1969 A new approach to the mapping of three-dimensional radionuclide distributions *J. Nucl. Med.* **10** 363
PATTON J A, BRILL A B and KING P H 1972 A new mode of collection and display of three-dimensional data for static and dynamic radiotracer studies *Medical Radioisotope Scintigraphy, Proc. Symp., Monte Carlo, October 1972* (Vienna: International Atomic Energy Agency) pp 355–68
PATTON J A, BRILL A B and KING P H 1974 Transverse section brain scanning with a multicrystal cylindrical imaging device *Tomographic Imaging in Nuclear Medicine* ed G S Freedman (New York: Society of Nuclear Medicine) pp 28–43

PEASE M 1946 A note on the radiography of paintings *Metropolitan Museum of Art Bull.* **4** 136–9

PELISSIER G 1931 Radiographie de face de la colonne cervicale dans son ensemble. Technique nouvelle *Bull. Soc. Radiol. Fr.* **19** 361

PETERS T M 1974 Spatial filtering to improve transverse tomography *IEEE Trans. Biomed. Eng.* **BME-21** 214–9

PICKENS D R 1977 The design, construction and preliminary testing of a mutually orthogonal coincident focal point tomographic brain scanner *MS Thesis* Vanderbilt University

PICKENS D R, KING P H, PATTON J A and BRILL A B 1978 The design, construction and preliminary testing of a mutually orthogonal coincident focal point tomographic scanner *Proc. 13th Annual Meeting of the Association for the Advancement of Medical Instrumentation, Arlington, Virginia* (Association for the Advancement of Medical Instrumentation)

PICKENS D R, SWEENEY J R and KING P H 1971 A feasibility study for a multidetector orthogonally coincident focal point tomographic scanner *Vanderbilt University Report*

PIERQUIN B 1961 La tomographie transversale: technique de routine en radiotherapy *J. Radiol. Électrol.* **42** 131–6

PIRIE A H 1911 A stereoscope for use with the fluorescent screen *Arch. Röntgen Ray* **16** 227–8

POHL E 1927–32 Verfahren und Vorrichtung zur roentgenphotographischen Wiedergabe eines Körperschnittes unter Ausschlub von davor und dahinter liegenden Teilen *German Patent* 544200

POHL E 1930–2a Verfahren und Vorrichtung zur radioskopischen und radiographischen Wiedergabe eines Körperschnittes unter Ausschlub von davor und dahinter liegenden Teilen *Swiss Patent* 155613

POHL E 1930–2b Improvements in and relating to the radioscopic and radiographic reproduction of a section through a body *UK Patent* 369662

PORTES F and CHAUSSE M 1921–2 Procédé pour la mise au point radiologique sur un plan sécant d'un solide, ainsi que pour la concentration sur une zone déterminée d'une action radiothérapeutique maximum, et dispositifs en permettant la réalisation *Fr. Patent* 541914 (NB Not 541941 as quoted by Andrews (1936) and others which is a patent for 'a machine for washing, drying and ironing linen')

PONTHUS P and MALVOISIN J 1937a Mémoires originaux; principe d'une méthode d'examen radioscopique 'en coupe' de l'organisme; la Stratiscopie *J. Radiol. Électrol.* **21** 337–43

PONTHUS P and MALVOISIN J 1937b Utilité de l'examen radioscopique 'en coupe' de l'organisme. Appareil schématique montrant la possibilité de cet examen *Bull. Mém. Soc. Radiol. Méd. Fr.* **25** 393–7

POWELL M D, MONAHAN W G and BEATTIE J W 1970 Gamma–gamma coincidence detector *Radiology* **94** 197

PRIOR K 1988 private communication (letter to S Webb, 11 March)

RADON J 1917 Über die Bestimmung von Functionen durch ihre Integralwerte längs gewisser Mannigfaltigkeiten (On the determination of functions from their integrals along certain manifolds) *Ber. Verbhandl. Sächs. Akad. Wiss. Leipzig, Math.-Phys. Kl.* **69** 262–77

RAMACHANDRAN G N and LAKSHMINARAYANAN A V 1971 Three-dimensional reconstruction from radiographs and electron micrographs: application of convolutions instead of Fourier Transforms *Proc. Natl. Acad. Sci. USA* **68** 2236–40

RANKOWITZ S, ROBERTSON J S, HIGINBOTHAM W A and ROSENBLUM M J 1962 Positron scanner for locating brain tumours *IRE Int. Conv. Record* Part 9 (New York: IRE) pp 49–56

RENANDER M A 1928 *Catalogue des Portraits des Membres du Deuxième Congres International de Radiologie, Stockholm, 23–27 July 1928, Acta Radiol., Suppl.,* **5**

REYNOLDS L 1945 The history of the use of the röntgen ray in warfare *Am. J. Röntgenol.* **54** 649–71

REYNOLDS R J 1961 The early history of radiology in Britain *Clin. Radiol.* **12** 136–42

RICHARDS A G 1966–70 Variable depth laminography with means for highlighting the detail of selected lamina *US Patent* 3499146

RICHARDS A G 1970–3 Method and apparatus for variable depth laminagraphy US Patent 3742236

RICHEY J B, McBRIDE T R and COVIC J 1977–9 Variable collimator *US Patent* 4143273

ROBERTSON J S and BOZZO S R 1964 Positron scanner for brain tumours *Proc. 6th IBM Medical Symp.* (Poughkeepsie: Brookhaven National, Laboratory) pp 631–45

ROBERTSON J S, MARR R B, ROSENBLUM M, RADEKA V and YAMAMOTO Y L 1974 32-crystal positron transverse section detector *Tomographic Imaging in Nuclear Medicine* ed G S Freedman (New York: Society of Nuclear Medicine) pp 142–53

ROBERTSON J S and NIELL A M 1962 Use of a digital computer in the development of a positron scanning procedure *Proc. 4th IBM Medical Symp.* (New York: Endicott) pp 77–103

ROBIN P A 1945 A simple apparatus for body section tomography *Mil. Surgeon* **96** 273–5

RONNEAUX G, DEGAND F, WATTEZ E and SAGET 1939 Essai de radio-photographie appliquée à la stratigraphie pulmonaire *Bull. Mém. Soc. Électrol. Radiol. Méd. Fr.* **26** 687–93

ROSS J A 1946 Stereoscopic screening *Br. J. Radiol.* **19** 156–7

ROWBOTTOM M and SUSSKIND C 1984 *Electricity and Medicine; History of Their Interaction* (San Francisco, California: San Francisco Press)

ROWLEY P D 1969 Quantitative interpretation of three-dimensional weakly refractive phase objects using holographic interferometry *J. Opt. Soc. Am.* **59** 1496–8

RUDIN S, RIDER K L and HART H E 1974 Gamma camera tomographic magnifying collimator systems *Tomographic Imaging in Nuclear Medicine* ed G S Freedman (New York: Society of Nuclear Medicine) pp 84–101

RUTHERFORD E 1919 The development of radiology: The First Mackenzie Davidson Memorial Lecture *Proc. R. Soc. Med.* **13** 147–57

SAKUMA S 1989 private communication (part translation of Takahashi obituary) (letter to S Webb, 25 May)

SANITAS 1934–9 Gerät zur röntgenographischen Darstellung einer planparallelen Körperschicht *German Patent 676594*

SANITAS 1935–8a Gerät zur röntgenographischen Darstellung von Körperschnitten *German Patent 665336*

SANITAS 1935–8b Gerät zur röntgenographischen Darstellung von Körperschnitten *German Patent 663476*

SCHMIDLIN P 1972 Three-dimensional scintigraphy with an Anger camera and a digital computer *Medical Radioisotope Scintigraphy, Proc. Symp., Monte Carlo, October 1972* (Vienna: International Atomic Energy Agency) pp 369–79

SCHONANDER G 1953 Design in radiological apparatus *Modern Trends in Diagnostic Radiology* ed J W McLaren (London: Butterworth) pp 1–14

SCOTT J 1940 A simple localiser *Radiography* **6** 143–5

SCOTT J 1960 Ancient and modern *Radiography* **26** 97–107

SIEMENS 1975–9 Apparatus for use in producing an image of a cross-section through a body *UK Patent 1546158*

SIRI W E 1949 Major organic metabolites *Isotopic Tracers and Nuclear Radiations with Applications to Biology and Medicine* (New York: McGraw-Hill) p 517

SNOW W 1939 Stereoröntgenoscopy using polarization *Am. J. Röntgenol.* **42** 143–4

SPEED K and BRACKIN R E 1955 Use of the tomogram after attempted joint fusion *Am. J. Surg.* **89** 872–4

STABLER E P 1974 Nuclear medicine tomography at Upstate Medical Centre NY *Tomographic Imaging in Nuclear Medicine* ed G S Freedman (New York: Society of Nuclear Medicine) pp 122–31

STAMM R W 1941 The polaroid stereoscope *Am. J. Röntgenol.* **45** 744–52

STEPHENSON J J 1950a Horizontal body section radiography *Br. J. Radiol.* **23** 319–34

STEPHENSON J J 1950b Horizontal body section radiography *Proc. 6th Int. Congr. on Radiology, London* p 80

STOCKING B 1979 X-rays highlight the doctor's dilemma *New Sci.* 11 January 84–7

STOCKING B and MORRISON S L 1978 *The Image and the Reality* (Oxford: OUP)

STODDART H F and STODDART H A 1979 A new development in single gamma transaxial tomography: Union Carbide focused collimator scanner *IEEE Trans. Nucl. Sci.* **NS-26** 2710–2

SWEET W M 1903 Locating foreign bodies by the Röntgen ray *Trans. of the American Röntgen Ray Society, 4th Ann. Meet., December 1903* pp 82–8

SWEET W H 1951 The uses of nuclear disintegrations in the diagnosis and treatment of brain tumour *New Engl. J. Med.* **245** 875–8

SWIFT R D 1980–1 Mechanical X-ray scanning *UK Patent* 2076250A

SWINDELL W 1970 A non-coherent optical analogue image processor *Appl. Opt.* **9** 2459–69

TAKAHASHI S 1957 *Rotation Radiography* (Tokyo: Japan Society for the Promotion of Science)

TAKAHASHI S 1965 Conformation radiotherapy: rotation techniques as applied to radiography and radiotherapy of cancer *Acta Radiol., Suppl.* 242

TAKAHASHI S 1969 *An atlas of Axial Transverse Tomography and Its Clinical Applications* (Berlin: Springer)

TAKAHASHI S 1985 Obituary *J. Clin. Sci.* **21** 1691–5 (in Japanese)

TAKAHASHI S and MATSUDA T 1960 Axial transverse laminagraphy applied to rotational therapy *Radiology* **74** 61–4

TAYLOR H K 1938 Planigraphic examination of the thorax in tuberculous individuals *Quart. Bull. Sea View Hospital* **3** 357–74

TAYLOR J H 1967 Two-dimensional brightness distributions of radio sources from Lunar occultation observations *Astrophys. J.* **150** 421–6

TAYLOR S 1976–7 Radiography with noble gas containing detector cells *US Patent* 4048503

TER POGOSSIAN M M 1985 Positron emission tomography instrumentation *Positron Emission Tomography* ed M Reivich and A Alavi (New York: A R Liss) pp 43–61

TER POGOSSIAN M M 1988 private communication (letter to S Webb, 17 October)

TER POGOSSIAN M M, FICKE D C, YAMAMOTO M and HOOD J T SR 1982 Super PETT 1: a positron emission tomograph utilising time-of-flight information *IEEE Trans. Med. Imag.* **MI-1** 179–87

TER POGOSSIAN M M and HERSCOVITCH P 1985 Radioactive oxygen-15 in the study of cerebral blood flow, blood volume and oxygen metabolism *Semin. Nucl. Med.* **15** 377–94

TETEL'BAUM S I 1956 About the problem of improvement of images obtained with the help of optical and analogue instruments *Bull. Kiev Polytech. Inst.* **21** 222

TETEL'BAUM S I 1957 About a method of obtaining volume images with the help of X-rays *Bull. Kiev Polytech. Inst.* **22** 154–60 (Engl. transl. by Professor J W Boag, Institute of Cancer Research, 1987)

329

THOMPSON C J, YAMAMOTO Y L and MEYER E 1978 Positome-2: a high efficiency PET device for dynamic studies *J. Comput.-Assisted Tomogr.* **2** 650–1

THOMSON E 1896 Stereoscopic Röntgen pictures *Electr. Eng.* **21** 256; *Electr. World* **27** 280

TOBB M 1950 Le Polytome de M M Sans et Porcher *J. Radiol. Électrol.* **31** 300–2

TOBIAS C A and ANGER H O 1953–7 Radioactivity distribution detector *US Patent 2779876*

TOMES SIR J A 1897 *A System for Dental Surgery* 4th edition, revised and enlarged by C S Tomes (Philadelphia; P Blackiston)

TRETIAK O J, EDEN M and SIMON W 1969 Internal structure from X-ray images *Proc. 8th Int. Conf. on Medical and Biological Engineering and 22nd Ann. Conf. on Engineering in Medicine and Biology, Chicago, Illinois, July 1969* ed J E Jacobs, Session 12-1

TROTT N G 1969 Mechanical scanning systems—rapporteur report in Radioactive isotopes in the localisation of tumours *Int. Nuclear Medicine Symp., London, September 1967* ed V R McCready, D M Taylor, N G Trott, C B Cameron, E O Field, R J French and R P Parker (London: Heinemann) pp 11–5

TWINING E W 1937a Tomography, by means of a simple attachment to the Potter–Bucky couch *Br. J. Radiol.* **10** 332–47

TWINING E W 1937b Discussion on the clinical value of the tomograph *Proc. R. Soc. Med.* **31** 386–96

VAINSTEIN B K 1970 Finding the structure of objects from projections *Kristallografiya* **15** 894–902 (Engl. transl. 1971 *Sov. Phys–Crystallogr.* **15** 781–7)

VALLEBONA A 1930a Ein technisher Vorgang zur Trennung der Schatten mittels Röntgenaufnahmen *Sanitärer Kong. der Zivilhospitale, Genua, 26 February 1930*

VALLEBONA A 1930b Una modalità di tecnica per la dissociazione radiografica delle ombre *Com. Congr. Sanit. Osped. Civ. Genova-Liguria Medica April 1930* vol 4

VALLEBONA A 1930c Una modalità di tecnica per la dissociazione radiografica della ombre *Radiol. Med.* **5** 629

VALLEBONA A 1930d Una modalità di tecnica per la dissociazione radiografica delle ombre applicata allo studio del cranio *Radiol. Med.* **17** 1090–7 (Engl. transl. Romano W L 1964 *Classic Descriptions in Diagnostic Röntgenology* ed A J Bruwer (Springfield, Illinois: Charles C Thomas) pp 1419–22)

VALLEBONA A 1930e Ein technisher Vorgang zur Trennung der Schatten bei Röntgenaufnahmen bei der Untersuchung des Schádels *Radiol. Med.* H **9**

VALLEBONA A 1930f Eine technische Modifikation fur die radiographische Trennung der Schatten am Schädel *Zentral bl. Radiol.* **9** 590

VALLEBONA A 1931 Radiography with great enlargement (microradiography) and a technical method for the radiographic dissociation of the shadow *Radiology* **17** 340—1

VALLEBONA A 1932a Bezüglich eines neuen radiographischen, mit 'Planigraphie' bezeichneten Verfahrens *Radiol. Med.* H **8**

VALLEBONA A 1932b A proposito di un nuovo metodo radiografico chiamato 'Planigrafia' *Radiol. Med.* **19** 869—71

VALLEBONA A 1933 Uber die Methoden zur Aufnahme von Röntgenbildern, die eine Zerlegung der Schatten ermöglichen *Fortschr. Geb. Röntgenstr.* **48** 599—605

VALLEBONA A 1934 Il cosidetto 'metodo planigrafico' *Radiol. Med.* **21** 58

VALLEBONA A 1948 I nuovi orizzonti della Stratigrafia nei varî campi della medicina. L'esplorazione stratigrafica tridimensionale *Inform. Med.* **2** 89—96

VALLEBONA A 1950a New applications of stratigraphy *Proc. 6th Int. Congr. on Radiology, London* p 89

VALLEBONA A 1950b Axial transverse laminography *Radiology* **55** 271—3

VALLEBONA A and BISTOLFI ST 1935a Bemerkungen zu dem Aufsatz 'Zur Tomographie' von Karol Mayer *Fortschr. Geb. Röntgenstr.* **52** 624

VALLEBONA A and BISTOLFI ST 1935b Uber die vershiedenen technischen Lösungen der Stratigraphie *Fortschr. Geb. Röntgenstr.* **52** 607—18

VALLEBONA A and BISTOLFI ST 1935c Erwiderung auf die Ausführungen von Herrn Dr Grossmann *Fortschr. Geb. Röntgenstr.* **52** 620—1

VALLEBONA A and BISTOLFI F 1973 Principles of physics *Modern Thin Section Tomography* ed A Berrett, S Brunner and G E Valvassori (Springfield, Illinois: Charles C Thomas) pp 18—27

VAN DER PLAATS G J 1932 Eenige opmerkingen over de zoogenaamde planigraphie met Röntgenstralen *Ned. Tijdschr. Geneesk* **76** 1081—6

VERSE H 1954—63 Apparatus for tomographic fluoroscopy with the use of image amplification *US Patent* 3091692

VIETEN H 1936—9 Verfahren zum röntgenographishen Darstellen eines Körperschnittes *German Patent* 672518

VISWANATHAN R and KESAVASWAMY P 1940 A simple method of tomography *Indian Med. Gaz.* **75** 279—82

VOGEL R A, KIRCH D, LEFREE M and STEEL P C 1978 A new method of multiplanar emission tomography using a seven pinhole collimator and an Anger scintillation camera *J. Nucl. Med.* **19** 648—54

VOGT O 1935 Kurze technische Erklärung des 'Tomograph' für Körper-Schnittbilder *Fortschr. Geb. Röntgenstr.* **52** 44—6

VUORINEN P 1959 The röntgenographic slit methods *Acta Radiol., Suppl.* 177

WAGNER W 1975—7 Device for measuring local radiation absorption in a body *US Patent* 4057725

WAGNER W 1976—9 Method of and device for measuring the distribution of radiation in a plane of a body *US Patent* 4144570

WALKER W G 1968—71 Tomographic radiation camera *US Patent* 3612865

WATSON W AND SONS LTD 1912 Correspondence *Arch. Röntgen Ray* **16** 369—70

WATSON W 1936—8 Improvements in or relating to X-ray apparatus *UK Patent* 480459

WATSON W 1937—9 Improvements in or relating to X-ray apparatus *UK Patent* 508381

WATSON W 1939—40 X ray apparatus *US Patent* 2196618

WATSON W 1939 Differential radiography 1 *Radiography* **5** 81—8

WATSON W 1940 Differential radiography 2 *Radiography* **6** 161—72

WATSON W 1943 Differential radiography 3 *Radiography* **9** 33—8

WATSON W 1947—8 Improvements in or relating to radiography *UK Patent* 601806

WATSON W 1950 Simultaneous multisection radiography *Proc. 6th Int. Congr. on Radiology, London* p 71

WATSON W 1951 Simultaneous multisection radiography *Radiography* **17** 221—8

WATSON W 1951—4 Improvements in or relating to apparatus for radiography and X-ray fluoroscopy *UK Patent* 705297

WATSON W 1953a Simultaneous multisection radiography *Modern Trends in Diagnostic Radiology* ed J W McLaren (London: Butterworth) pp 15—25

WATSON W 1953b Layer fluoroscopy *Radiography* **19** 189—96

WATSON W 1956 Some observations on stratigraphic technique *Stratigrafia* **1** 208—21

WATSON W 1957 Tomoscopy *Stratigrafia* **2** 207—15

WATSON W 1961 1895 and all that *Radiography* **27** 305—15

WATSON W 1962 Axial transverse tomography *Radiography* **28** 179—89

WEBB S 1988 In the beginning *The Physics of Medical Imaging* ed S Webb (Bristol: Adam Hilger) pp 7—19

WEBER R L 1973 *A Random Walk in Science* (Bristol: Adam Hilger) (quoting letter from Sir Isaac Newton to Sir Robert Hooke (5 February 1676))

WEINBREN M 1946 *A Manual of Tomography* (London: H K Lewis)

WEIR J and ABRAHAMS P 1978 *An Atlas of Radiological Anatomy* (Tunbridge Wells, Kent: Pitman Medical) pp 66, 272

WESTRA D 1966 *Zonography; the Narrow Angle Tomography* (Amsterdam: Excerpta Medica)

WESTRA D 1972 Laudation to Prof Dr B G Ziedses des Plantes *Radiol.*

Clin. Biol. **41** 326–33

WESTRA D 1973 History of tomography *Modern Thin Section Tomography* ed A Berrett, S Brunner and G E Valvassori (Springfield, Illinois: Charles C Thomas) pp 3–17

WHEATSTONE C 1838 Contributions to the physiology of vision—part the first. On some remarkable, and hitherto unobserved phenomena of binocular vision *Phil. Trans. R. Soc. London* Part 1 371–94 (reprinted 1964 *Classic Descriptions in Diagnostic Röntgenology* ed A J Bruwer (Springfield, Illinois: Charles C Thomas) pp 990–1010)

WHEATSTONE SIR C 1879 *The Scientific papers of Sir C Wheatstone* (reprinted from the *Philosophical Transactions* of 1838) (London: Physical Society)

WHEELER D and SPENCER E W 1940 Simplified planigraphy *Radiology* **34** 499–502

WILSON R 1911 Correspondence *Arch. Röntgen Ray* **16** 280

WRENN F R, GOOD M L and HANDLER P 1951 The use of positron emitting radioisotopes for the localization of brain tumours *Science* **113** 525–7

YAMAMOTO Y L 1988 private communication (letter to S Webb, 11 October)

YAMAMOTO Y L, THOMPSON C J and MEYER E 1976 Evaluation of positron emission tomography for study of cerebral hemodynamics in a cross-section of the head using positron emitting gallium 68- EDTA and krypton 77 *J. Nucl. Med.* **17** 546

YAMAMOTO Y L, THOMPSON C J, MEYER E, ROBERTSON J S and FEINDEL W 1977 Dynamic positron emission tomography for study of cerebral hemodynamics in a cross-section of the head using positron-emitting ^{68}Ga-EDTA and ^{77}Kr *J. Comput.-Assisted Tomogr.* **1** 43–56

ZIEDSES DES PLANTES B G 1931 Een bijzondere methode voor het maken van Röntgenphotos van schedel en wervelkolom *Ned. Tijdschr. Geneesk.* **75** 5218–22 (Eng. transl. de Ruyter H A 1964 *Classic Descriptions in Diagnostic Röntgenology* ed A J Bruwer (Springfield, Illinois: Charles C Thomas) pp 1424–30) (another Engl. transl. Ziedses des Plantes 1973 *Selected works of B G Ziedses des Plantes* (Amsterdam: Excerpta Medica) pp 1–7)

ZIEDSES DES PLANTES B G 1932 Eine neue methode zur differenzierung in der Röntgenographie (Planigraphie) *Acta Radiol.* **13** 182–92

ZIEDSES DES PLANTES B G 1933 Planigraphie *Fortschr. Geb. Röntgenstr.* **47** 407–11

ZIEDSES DES PLANTES B G 1934a Planigraphy, a röntgenographic differentiation method *Doctoral Thesis* Utrecht (Utrecht: Kemmink) (Engl. transl. Ziedses des Plantes 1973 *Selected works of B G Ziedses des Plantes*

(Amsterdam: Excerpta Medica) pp 9–84)

ZIEDSES DES PLANTES B G 1934b Planigraphie: une methode permettant en radiographie d'obtenir une image nette de la section d'un objet à un plan bien déterminé *J. Radiol. Électrol.* **18** 73–6

ZIEDSES DES PLANTES B G 1934c Il cosidetto 'metodo planigrafico' *Radiol. Med.* **21** 56–8

ZIEDSES DES PLANTES B G 1935 Seriescopy, Een röntgenographische methode welke het mogelijk maakt achtereenvolgens een oneindig aantal evenwijdige vlakken van het te onderzoeken voorwerp afzonderlijk te beschouwen *Ned. Tijdschr. Geneesk.* **51** 5852–6

ZIEDSES DES PLANTES B G 1936 Röntgenologic method and apparatus for consecutively observing a plurality of planes of an object *UK Patent* 487389

ZIEDSES DES PLANTES B G 1937 Over analytische röntgenografie *Handelingen van het XXVIe Nederlandsch Natuuren Geneeskundig Congr., Utrecht, March 1937*

ZIEDSES DES PLANTES B G 1938 Seriescopy: Eine röntgenographische Methode welche ermöglicht, mit Hilfe eineger Aufnahmen eine unendliche Reihe paralleler Ebenen in Reichenfolge gesondert zu betrachten *Fortschr. Geb. Röntgenstr.* **57** 605–19 (Engl. transl. Ziedses des Plantes 1973 *Selected Works of B G Ziedses des Plantes* (Amsterdam: Excerpta Medica) pp 129–43)

ZIEDSES DES PLANTES B G 1950a Direct and indirect autoradiography. *Proc. 6th Int. Congr. of Radiology, London* p 172, paper B24 (abstract) (reprinted in full Ziedses des Plantes 1973 *Selected works of B G Ziedses des Plantes* (Amsterdam: Excerpta Medica) pp 199–204)

ZIEDSES DES PLANTES B G 1950b Direct and indirect radiography (reprint Ziedses des Plantes 1973 *Selected works of B G Ziedses des Plantes* (Amsterdam: Excerpta Medica) pp 199–204)

ZIEDSES DES PLANTES B G 1964 Geometrische Probleme der Tomographie *Röntgenblatter* **17** 357–68

ZIEDSES DES PLANTES B G 1971 Body section radiography; history, image formation, various techniques and results *Australas. Radiol.* **15** 57–67

ZIEDSES DES PLANTES B G 1973 *Selected works of B G Ziedses des Plantes* (Amsterdam: Excerpta Medica)

ZIEDSES DES PLANTES B G 1978 Bevorrugte röntgendarstellung bestimmter teile de menseklichen korpers in vergangunlich und gegenwert *Röntgenstrahlen* **38** 40–50

ZIEDSES DES PLANTES B G 1988a private communication (letters to S Webb, 14 April and 25 May)

ZIEDSES DES PLANTES B G 1988b Il lavoro creativo è privilegio degli esperti?; Is creative work the privilege of experts? *Riv. Neuroradiol.* **1** vol 1 (Udine: Arti Grafiche Friolane) pp 9–16

ZINTHEO C J 1939 Planigraphy *X-ray Technician* March 206–11

Index

Numbers in **bold** refer to **figures** and those in *italics* refer to *tables*.

Printed and bound by CPI Group (UK) Ltd, Croydon, CR0 4YY

17/10/2024

01775690-0014